CAUSE, EXPERIMENT *and* SCIENCE

Stillman Drake

CAUSE, EXPERIMENT

and SCIENCE

A Galilean dialogue incorporating
a new English translation of Galileo's
"Bodies That Stay atop Water,
or Move in It"

The University of Chicago Press / Chicago and London

The University of Chicago Press, Chicago 60637
The University of Chicago Press, Ltd., London

Printed in the United States of America
85 84 83 82 81 5 4 3 2 1

Library of Congress Cataloging in Publication Data

Drake, Stillman,
Cause, experiment, and science.

Includes index.
1. Hydrostatics. I. Galilei, Galileo, 1564–1642.
Discorso al serenissimo don Cosimo II, gran duca di Toscana, intorno alle cose, che stanno in su l'acqua.
English. 1981. II. Title
QC147.D7 532'.2 81-2974
ISBN 0-226-16228-1 AACR2

STILLMAN DRAKE is emeritus professor of the history of science at the University of Toronto and author of many books on Galileo, including *Galileo at Work: His Scientific Biography*.

To my sons MARK *and* DANIEL

Contents

Preface

The search for causes of events in nature guided science from the time of Aristotle to that of Galileo. The dawn of modern science saw this search superseded by a quest for laws of nature based on experiment and measurement. One of the first printed books to herald that new approach is here presented in modern English. It was followed by a revolution in physics that started with Galileo's *Two New Sciences* and matured in Newton's *Mathematical Principles of Natural Philosophy*.

In his old age Galileo adopted the dialogue form as the best way to present new scientific ideas to the educated general public, not only to reduce readers' boredom but also to make it clear that there were reasons on both sides and that the task of science was to judge preponderance of evidence rather than to discover final truth. For similar reasons I have adopted the dialogue form in presenting Galileo's first book on experimental physics. Three friends discuss Galileo's views, one offering further information and support, another raising understanding questions and suggestions, and the third defending the traditional approach to physics, or natural philosophy as it was then called. The interlocutors are those introduced long ago in Galileo's last two books, and I have placed in their mouths speeches that I think are in character with the arguments Galileo assigned them in other debates on cosmology, motion, and mechanics. Here the topic is hydrostatics, seemingly a matter of little interest, in which even dialogue might not save readers from boredom.

My purpose is to throw light on the sudden rise and spread of general as well as professional interest in experimental science during the seventeenth century. Various explanations have been previously offered in terms of philosophical trends and of sociological forces. Such accounts interest scholars and satisfy those historians of science who are concerned principally with great advances made by a handful of celebrated early scientists. They do little to explain the attraction such achievements held for a wider public. Galileo's book on hydrostatics sold out so fast that an expanded second edition was called for a month after the first edition was printed. The reasons why an apparently dull topic met with extraordinary public interest may afford us a new key to the scientific revolution of the seventeenth century.

What was at stake was not just the floating of solid bodies on water, but a whole new approach to the study of nature. Galileo's adversaries, who were professors of philosophy, recognized this at once, and a long and heated controversy at Florence preceded his publication of the *Discourse* in 1612. He carried the fight to general readers by writing his book in Italian rather than in the Latin preferred by professors. That the public was no slower than philosophers to recognize the real issue is shown by the book's prompt best-seller status and by the number of published attacks that followed—likewise printed in Italian to reach those who might have been swayed by Galileo to challenge the official natural philosophy of the universities.

The real issue was an assumption, long accepted without question by bookish people, that a world on paper was superior to the world of actual experience. Late Renaissance Italy, and Florence in particular, was ready to challenge this assumption. In physical science, Galileo's discourse on bodies that float or move in water put the means of a successful step beyond Aristotelian tradition into the hands of any intelligent reader, by replacing causal speculation with deliberately designed experiments.

Few historians of science today credit experiment with any great role in Galileo's science. That is because the opening battle over the concept of cause best employed in science has been neglected in favor of Galileo's more famous battle a quarter-century later. The superior interest of Copernicanism has naturally tended to distract attention from the first attempt by Aristotelians to stifle the infant experimental science in its cradle. You will see here how Galileo first linked mathematics to observation and why people who were not mathematicians saw the value (and the fun) of doing that. This is of no little use in understanding his later and more important book on physics, which sold so poorly that Galileo's publishers abandoned their project of reprinting his collected works when he was old and blind.

Into the fictional dialogue in which I have embedded my new translation, I have introduced some matters that Galileo's friends would probably not have discussed. Various things they took for granted from their university courses in philosophy would not be known to many modern readers, so I have smuggled them into the conversations. More modern questions, such as semantic issues that would have escaped people at the time, are likewise brought in. Such anachronisms are deliberate; except for them, I have adhered mainly to information and arguments for which there is direct or indirect contemporary documentation. Sources are identified in the Appendix, where some further historical information is also given.

Simplicio is intended to speak for the philosophers who opposed

Galileo's science, as he does in Galileo's own dialogues, but he stands especially for Cesare Cremonini, who was Galileo's personal friend and scientific adversary at Padua.

Giovanfrancesco Sagredo, born at Venice in 1571, studied with Galileo about 1597 and remained one of his closest friends until his death in 1620. Active in government affairs, he was also a talented amateur scientist who performed many experiments in optics and thermometry. His reports concerning hydrostatic experiments here include my own findings, obtained by using paraffin wax weighted with bits of copper wire. He is intended to speak for the educated layman of the period, interested in nature and disenchanted with empty verbalisms.

Filippo Salviati, born at Florence in 1582, invited Galileo to join a circle of philosophers and literary men who met frequently at his palazzo, after Galileo returned to Florence in 1610. It was there that the controversy over floating bodies began. Galileo also frequently stayed with Salviati at his villa near Florence, where he wrote most of his *Discourse* and his *Sunspot Letters.* Late in 1613 Salviati went to Spain, embarking from Venice. Although he visited Cremonini at Padua, he probably never actually met Sagredo, who was frequently absent from Venice on government missions. Salviati is supposed to speak for Galileo as well as for himself, being personally familiar with and favorable to Galileo's scientific conceptions.

The translation of Galileo's *Discourse* is set in type distinguishable from that used for fictional dialogue, so that it may easily be read consecutively without my informal dialogue commentary. It follows the text as edited by Antonio Favaro in the fourth volume of the Edizione Nazionale of Galileo's collected works, page numbers of which are shown in brackets beside the translation. An earlier English translation, by Thomas Salusbury, was published at London in 1665, but most copies were destroyed in the Great Fire. It was reprinted in facsimile in 1960 with my notes and introduction, but it is difficult to read because of the antique style and the many errors of both translator and printer. A second facsimile reprint, without notes or corrections, appeared in London in 1968 as part of Salusbury's collected translations. I hope that this new version, in modern English with commentary, will bring to general attention many interesting facets of Galileo's first contribution to experimental science that have gone unnoticed. A brief history of hydrostatics to the time of Galileo and of the bearing of his *Discourse* on the ensuing conflict with theologians is given in the Appendix. At its end are some further commentaries whose presence is indicated in the text by a dagger (†) at the earliest relevant passage.

My heartfelt thanks go to Miss Beverly Jahnke for preparing the typescript and to Dr. Charles V. Jones for his thoughtful reading and his helpful comments on it. For every other aid and comfort to me during the work I am obliged—infinitely obliged—to Florence Selvin Drake.

GALILAEUS GALILAEI PATRICIUS FLOR.
AET. SUAE
ANNUM AGENS QUADRAGESIMUM

Introduction

Galileo acquired international celebrity in 1610, when he announced some startling telescopic discoveries in a small book called *The Starry Messenger.* Rather surprisingly, considering the common belief that Galileo was a Copernican zealot from his early years, the new astronomy was hardly mentioned. Up to 1610, when Galileo was forty-six years old, most of his attention had been directed toward problems of mechanics and the motions of heavy bodies. Until he turned his telescope to the heavens, he remained a physicist rather than an astronomer.

His sudden fame enabled him to obtain the post of chief mathematician and philosopher to the grand duke of Tuscany. Resigning his professorship of mathematics at the University of Padua, where he had lived since 1592, he returned to Florence in September

1610, freed from active teaching duties and with the intention of publishing a number of books based on his researches in physics. The title "philosopher" was meant to indicate what we would call "physicist," physics then being included in natural philosophy and not generally recognized as a mathematical discipline. Galileo intended to covert it into one, despite a certain prejudice against that approach on the part of Aristotelian professors of philosophy, who then dominated university education.

As it happened, the focus of attention in Galileo's first book after his move to Florence was on phenomena of floating in water. That was a subject on which Aristotle had written very little indeed, so there may be now no apparent reason why professors of philosophy should become upset about anything Galileo might say concerning it. In fact, however, they opposed him on this even more vigorously than they had attacked his recent telescopic discoveries. When he published his account of the matter and his answers to philosophers who had opposed him, they replied in no less than four published books, and the same philosophers enlisted support from theologians when the dormant Copernican battle began to take shape at the end of 1613.

In order to understand the sudden interest of professors of philosophy in matters of hydrostatics, it is helpful to know how the argument began. Although it is evident from what followed that Galileo had given much thought to the behavior of solids placed in water, he had not previously made an issue of it, and he did not start the fight by raising that issue for its own sake. During a conversation at the home of Filippo Salviati in the summer of 1611, the topic of condensation and rarefaction of matter was under discussion by some philosophers. This was a subject of considerable importance in Aristotelian physics, because the most natural explanation of it had been given by ancient Greek atomists, whose doctrine Aristotle opposed on every issue. In the discussion, ice was mentioned as an example of condensation, it being supposed that cold had the power to condense water. Since water was one of the four Aristotelian "elements," and cold was one of Aristotle's four explanatory properties (while moisture was another), the whole structure of Aristotelian natural philosophy was in a way brought into question if the concept of ice as water condensed by cold were successfully challenged. It was metaphysics rather than the behavior of solids in water that was under discussion when Galileo innocently dropped his bombshell into a philosophical disquisition.

"I should have thought that ice might rather be called 'rarefied water,'" remarked Galileo, "since it floats on water." The professors regarded this as an impertinence, thinking that no one could seriously regard a fluid as a condensed solid. In the case of water, however, Galileo did; melted ice takes up less room than the ice. Moreover, the hardness of ice is not proof of density, or one would have to say that lead is harder than steel, which it obviously is not. The philosophers accordingly returned to the matter of floating and asserted that ice floats not because it is lighter than water, but because of its flat shape, which prevents its piercing water's resistance to division. Galileo replied that a flat piece of ice shoved to the bottom of a tub of water will rise to the top, easily piercing the assumed resistance to division, if water has any such resistance—even with gravity acting to pull the ice downward. Hence he questioned any such "resistance."

The now disturbed philosophers pointed out to Galileo that water certainly does resist motion, for a sword that will easily cut through water edgewise is so stoutly resisted when struck flat on water that the swordsman's wrist may be injured. Such experiences proved to their satisfaction that the flatness of ice explained its floating. Galileo replied that he did not deny that water could resist *speed* of motion, but only that it resisted motion as such, since it is evident that a sword will sink in water in any position. Moreover, particles of dirt too small to be seen individually will settle out of muddy water if it is allowed to stand long enough. Those particles must weigh very little indeed; yet water offers no resistance sufficient to stop their sinking.

The original debate went no further, and the whole affair might have stopped there had it not been for an incident three days later.

It is important to remember that what was under discussion was a certain philosophical issue on which the professors had no doubt that they were entirely correct. Against this had been raised a terminological objection that seemed to them easy to dispose of as something no one with common sense would defend. Galileo, however, defended it quite seriously, answering each of the experiences they adduced in its support and adducing other experiences in favor of his position. Aristotle, whose physics they were defending, favored appeal to experience, so they could not object to that. Hence there was a real puzzle for the professors. They thoroughly understood the four elements and the four qualities (heat, cold, moisture, and dryness) governing their transformations; moreover, they knew the causes of natural motion (gravity and levity). Materials, forms or qualities, and causes were all there were to know in order to have perfect science—that is, knowledge of things in terms of their causes, or knowledge of the "reasoned fact," as Aristotle defined science. Galileo's arguments must therefore be specious, even if his fallacies were not detectable at first sight. Plenty of other plausible fallacies had been exposed by philosophers in the past, and there was no reason to doubt that some fallacy would be detected in Galileo's position.

The attitude of professors of philosophy at this stage of the debate was that of a responsible parent toward a refractory child who has got hold of some idea that he imagines can refute the wisdom of the ages. It was inconceivable to them that Aristotelian natural philosophy, after centuries of study and refinement, might be mistaken through and through. Nor did it occur to them that a small part of it, to which they had previously paid little attention, could be revised without affecting the entire causal structure. That was at once the strength and the weakness of the natural philosophy they had inherited; it was capable of explaining absolutely everything in nature, but if any flaw were found in it, vast sections necessarily came into doubt. Confident that they could not have been wrong, they saw their task only as finding and exposing some error in Galileo's position, which was based on the analysis of bodies in water given by Archimedes and which amounted to saying that the sole cause of floating or sinking in water was the lesser or greater density of the object relative to the density of water. A single exception to that rule would suffice to destroy Galileo's position, so they felt no need to reexamine their own statements.

In the course of discussing the problem with others, one of the professors talked with Lodovico delle Colombe, a Florentine with an old grudge against Galileo who was only too happy to assist. Colombe said that he would easily show by experiment that floating was not just a matter of relative densities, but that shape played an essential role in it. A flat ebony chip can be floated, whereas an ebony ball of the same weight cannot. That alone would serve to defeat Galileo and vindicate the philosophers.

When Galileo was told of this, he at first objected that it was a quibble; the original debate concerned bodies placed in water, and ebony truly *in* water would sink, no matter what its shape, just as ice placed in water would rise, whatever its shape. But he soon agreed to confront Colombe and debate any experiments offered, including experiments of his own. Before the meeting took place, however, the grand duke intervened. It was unseemly for his court mathematician and physicist to engage in noisy disputes; rather, Galileo should write out his position and his arguments, as he proceeded to do. Colome took Galileo's refusal to argue publicly as an admission of defeat and an unwillingness to submit to the authority of experiments against his fallacious reasoning.

Galileo's procedure was quite different from that of his adversaries, because their alleged causes could not be refuted by one test. He was as confident of the correctness of his position in the original debate as they were of theirs, but he felt no need to consult others about ways of defending it. Nothing he believed was threatened by any experiences yet adduced. Floating was not, to him, connected with a grand system of causal explanation for everything in the universe. Galileo's stake in the argument was small compared with the stake of the professors, who were obliged to know everything and to be right about everything. When applying for his court position, Galileo listed a dozen books he had in preparation, but hydrostatics was not included among the subjects he named. There is no reason to think that Galileo regarded the issue as of any great moment, or that he was belligerent about it. He simply did not want to be bothered with quibbles; he had other important work in progress that required nightly observations and innumerable time-consuming calculations by day, relating to the prediction of positions of Jupiter's satellites.

Nevertheless, the behavior of Colombe's ebony chip intrigued Galileo. It was clearly true that under some conditions a body denser than water could be floated, and indeed Aristotle had mentioned that fact in one brief chapter from which the professors had got their idea that water resists division. The floating of ebony chips *did* seem to contradict the principle of Archimedes, and, though Galileo appears not to have mentioned Archimedes during the original debate, which was over philosophical terminology and a mistaken conception of Aristotle's, he no more doubted that Archimedes was right about floating than the philosophers doubted Aristotle. Hence there was a similarity between the predicament of the professors after the initial discussion and that of Galileo after Colombe brought up the ebony chip. In both cases a puzzle existed that needed to be resolved, and there was equal confidence on both sides that its resolution would vindicate the position already taken on the floating of ice. It is instructive to compare the two approaches to a genuinely felt puzzle, for therein may be seen a departure of modern physics from traditional natural philosophy.

Galileo did not consult other scientists, mainly because no one in Florence knew more about floating than he did, but partly because the only help he could get from other scientists would have had its ultimate source in consulting nature itself in the matter of floating bodies denser than water. Nature was equally open to consultation by Galileo himself, so he closely observed the floating of such bodies. At once he noticed something that had previously escaped attention. That was why he agreed to debate with Colombe after his first refusal on the grounds that the ebony chip was a quibble.

What Galileo noticed was that bodies denser than water, when they do float, float under water, so to speak. Ordinary floating bodies have parts that project above the surrounding water surface. In the case of ice that was obvious from the fact (already mentioned) that melted ice takes up less room than the ice. Bodies denser than water, when they float, sink some distance below the water level before they come to rest. Having seen this, Galileo realized immediately that, far from violating the principle of Archimedes, such bodies obey it in a particularly interesting way. How he developed this will be seen in his book.

Thus the Aristotelian principle of appealing to experience had degenerated among philosophers into dependence on reasoning supported by casual examples and the refutation of opponents by pointing to apparent exceptions not carefully examined. Galileo took the basic principle seriously and thereby gained a fact of experience unknown to his adversaries that turned their own experiment against them. For, while it is possible to reconcile the experiment with the Archimedean principle, it cannot be explained by resistance of water to division in the sense intended by Aristotelians.

Why is it impossible to explain the floating of an ebony chip by resistance of water to division? Obviously that did not appear impossible to Galileo's opponents; on the contrary, they considered this the only way to explain the observed phenomenon, and they regarded Galileo as unrealistic and stubborn when he refused to grant to water any resistance whatever to division. In our superior knowledge of science, we are likely to miss the essential point of the dispute by judging Galileo's position in this matter incorrect and merely obstinate, as his adversaries did. For we now know that ordinary fluids at ordinary temperatures have certain viscosities, which differ greatly from one substance to another. We would not expect an ebony chip placed on molasses to sink through it to the bottom, or even a chip of lead, which is certainly denser than molasses. The viscosity of water may be extremely slight, but it exists; hence we may be tempted to suppose that it could be sufficient to restrain the descent of some body only a little denser than water, placed gently on its surface, forgetting that it would then be equally restrained below the surface. That is more or less the same mistake made by the philosophers but not by Galileo. It is important to see why they made it and why Galileo did not. After that it is possible to see why the fact that water at room temperature has a very small but measurable viscosity does not make Galileo wrong when he denied, in 1611, that water had any resistance at all to division.

The philosophers were absolutely certain that water resists division because it was a major assumption throughout Aristotle's physics that motion could not occur in an absolute void—that a medium capable of opposing motion must be present for motion to occur at all. It is of no importance in the present matter to know why Aristotle maintained this or how he purported to demonstrate its truth. What is important is that in Aristotle's physics countless implications of this assumption were scattered throughout his conclusions. That fact, not a knowledge of the viscosity of fluids, made Galileo's adversaries impervious to any of the experiments or reasons he adduced to refute the notion of fluid resistance to division. We could say that the Aristotelians felt threatened by Galileo's experiments and arguments, but that is a figure of speech, much like saying that Moscow feels threatened by the activities of dissident artists in the Soviet Union. The Aristotelians did not feel threatened at all; they simply knew that they were correct and that Galileo must be put right. It was their duty, as philosophers, to expose his error, and as soon as Colombe provided an experiment sufficient for the purpose, they felt they had done their duty. Looking back at the experiences and arguments they had adduced at the start, which were dreadfully inadequate and were immediately held up to scorn by Galileo, we see that only supreme confidence in their position can reasonably account for the sloppiness with which they supported it in a matter to which they had never given any thought, against an opponent who obviously had thought a great deal about it before he ventured to speak of ice as rarefied water, that being the kind of statement no one makes without having first prepared to defend it. It is a common mistake to suppose that the philosophers of Galileo's day were stupid or incompetent; they simply were not scientists and had no regard for scientists, as can be true even today.

Resistance of water to division, then, was asserted by the Aristotelians not on the basis of experimental evidence but according to a metaphysical principle that motion requires an opposing medium in order to occur at all. They were confident of this because an entire physics consistent with it had been constructed. It was far from evident that another physics, equally consistent, could be constructed without this principle, and even if one could they saw no reason to think it would be better, or even worth the trouble. This, then, is why they made the mistake of assuming that resistance of water to division causes ice and other flat objects to float on water.

Galileo did not make the same mistake because he recognized that many implications of resistance of water to division contradicted experience. The settling out of individually invisible particles from turbid water, given enough time, was one of these. If his opponents had felt obliged to assign a quantitative *measure* to the supposed resistance of water to division, they would have perceived the contradictions into which the assumption was bound to lead them. Against any such necessity, however, they had a general Aristotelian protection: it is finicky, in physics, to trouble about mathematical precision, which is not to be expected in material things. Hence none of Galileo's mathematical arguments touched the Aristotelian position at all, and it would be a waste of time for him to adduce them against philosophers. That is one of the reasons why Galileo wrote his *Discourse* in Italian for men of intelligence and common sense, rather than in Latin, like his *Starry Messenger,* for the information of philosophers.

It should now be clear why I said Galileo was not wrong in 1611 to say that water had *no* resistance to division, even though we now know that water has measurable viscosity. Certainly no one will say that the viscosity of water was detectable in Galileo's day, and thus neither he nor his adversaries could have taken it scientifically into account in explaining the behavior of solids in or on water. To understand the history of science, and to judge the scientific status of arguments at any given epoch, we must consider what was known at the time, what was not known, and what could not have been detected by any available means. It is then sometimes interesting to consider, as I shall do next, the reasons why a scientist formed some conclusion that later turned out to be wrong.

Galileo assumed that if water resisted division, such resistance would be the same at the surface as anywhere else in the water. I used the same assumption a moment ago, since it would be very hard to say whether or not it was simply wrong. But conditions at the surface of water are not the same as conditions within it, because of what has long been called "surface tension." (We go on using that term, invented in the eighteenth century, though it is a misnomer for a certain potential energy affecting a surface layer only a few molecules thick.) The anomalous floating pointed out by Colombe was later explained by surface tension. It would be unreasonable to blame Galileo for not thinking of that explanation, but we might wonder why he did not simply assume that conditions are different at the surface of water and within it. Of course he did know that an ebony chip behaves differently on the surface of water and within it, since that is precisely what was under debate. He might have explained that by assuming a kind of "skin resistance" of water to division, so I have had Sagredo and Salviati bring this up in my fictional dialogue. We will see that Galileo preferred to discuss an observed fact about the chip in the two placements; namely, that part of it remains dry in the floating position only. Some may not like the way he attempts to utilize this observed fact in his explanation, but it is worth noticing that in science he preferred observed facts to arbitrary assumptions.

If we could ask Galileo why he did not assume "skin resistance" of water, I believe he would reply that useful science had been held back for centuries by just that kind of ascription to matter of "occult properties," selected ad hoc to explain whatever phenomenon one pleased. Molière poked fun at such explanations later on by having a medical student reply, to the question why opium causes sleep, "because opium possesses the soporific quality." What had become a laughingstock by Molière's time was a respected form of explanation from the beginning of universities in the thirteenth century to the time of Galileo. Thus it was perfectly acceptable to say that a piece of lead falls to earth because lead possesses gravity. It is not always easy to tell at first whether statements of this form convey useful information or remain empty words. When *gravity* was still merely a word for *heaviness,* as it remained until Newton discovered the law of universal gravitation, the latter statement conveyed no more information than did Molière's spoof. But when *gravity* came to mean acceleration according to a definite law, every statement containing the word took on scientific significance. For similar reasons, the search for causes generally gave way in science to a quest for laws of nature, while the concept of cause passed back to the exclusive jurisdiction of philosophers, scientific statements being always capable of formulation without it.

The process by which causes gave way to laws in science may be considered as having begun when the word *cause* was first sharply defined for use in scientific inquiries. I contend that this happened when Galileo defined it in the hydrostatics controversy, though that may at first seem a very rash statement. The topic is so interesting and important that I shall discuss it at some length, remarking at the outset that my purpose is not to establish some absolute priority for Galileo in a matter in which no priority can perhaps be established at all, but rather to signal a situation that was, for practical purposes, the opening of the scientific revolution and yet has remained unnoticed and unexamined, so far as I know.

Because Aristotle defined science (*episteme*) as knowledge of things in terms of their causes, and because he was aware of the importance of definitions, one might suppose that he must also have defined cause (*aition*). That, however, seems not to have been the case. He did classify four types of cause, later given names by medieval commentators (material, formal, efficient, and final causes), but *cause* itself remained undefined. In saying this, I rely on my understanding of *Causality and Scientific Explanation,* a two-volume survey published in 1972 by William A. Wallace, who gave the traditional meaning of *cause* as "something that exists outside the mind and serves to explain, not merely in a logical way but in an ontological way, why the thing is as it is." Thus the notion of external existence of causes was included in any reference to them, though the determination of cause was a matter of reason rather than of direct observation.

Galileo, who like everyone else of his time was schooled in Aristotelian natural philosophy, expressed this in his first treatise on motion, written about 1591–92: "what we seek are the causes of effects, and these causes are not given to us by experience." Aristotle had made motion basic to all physics and, though Galileo disputed many of Aristotle's conclusions in this treatise, he did not depart from Aristotle's conception of science, whose task was to find the causes of effects, the effects being perceptible by the senses and their causes discoverable only by reason.

During the next twenty years Galileo was occupied mainly in studies of mechanics and of motion. In 1604 he discovered the law of falling bodies, and in 1608 he found that projectiles travel in parabolic paths, both discoveries having resulted from careful experimental measurements. In 1604–5 he disputed publicly with Cremonini (Simplicio in my dialogue) over the location of a new star, which the philosopher placed beneath the moon while the astronomer put it at least as far away as the outermost planets. Because the star was new, philosophers argued, it could not be made of the unchanging material of the heavens, and if made of elemental matter it must be situated in the elemental region below the moon. That was the "material cause" of its low place. Galileo said that it was poor judgment to abandon the senses and go searching for reason. Observation showed that the new star had no measurable parallax, and that sufficed to place it at an enormous distance. Thus, considerably before 1611, Galileo had come to attach more importance to experience in deciding scientific issues than he had been taught to do, though that does not mean that he attached any less importance to reason. Reason unsupported by direct experience, however, had led philosophers to some dubious conclusions via causal reasoning, both in matters of motion and in questions of celestial position. Having departed from them in those inquiries, it is not surprising that Galileo went still further in the controversy over bodies in water.

Colombe's intended refutation of Galileo's Archimedean position was based on the idea that shape was causally involved in the floating of ebony chips and the sinking of ebony balls. Preparing for debate against Colombe, Galileo set down this definition of cause: "That which, given, the effect is there; and, removed, the effect is taken away." His purpose in devising a definition of cause is evident, since the challenge meant a contest of experimental evidence relating to a causal conclusion. And since science consisted of causal explanation, in the view of Galileo as well as his opponents in 1611, his definition of cause reflected his conception of science and the role played by experiment in it.

Galileo's definition of *cause* for the purposes of scientific inquiry was included in the 1612 *Discourse* and appears not to have been challenged by Colombe or the professors of philosophy who wrote against that book, in which they did challenge nearly everything else. It would in fact be difficult to discredit Galileo's definition, with fits very well with common treatment of the cause-and-effect relationship in everyday language. Galileo probably had no ulterior motive in wording his definition as he did; it is probable that he merely fleshed out the aphorism for curing diseases that he must have learned as a medical student: *tollet causam,* "remove the cause." But since the occasion for his defining cause was connected with experiments, Galileo tended to think of the presence and absence of cause and effect in more literal and less metaphorical terms than had been usual, as if causes could be revealed to us by our senses. He did not repeat his definition in later books, but he retained it as a criterion in science when he wrote in his *Dialogue* of 1632: "Consider what there is that is new . . . and therein necessarily lies the cause of the new effect."

Twenty years and more elapsed between the *Discourse* and Galileo's two most famous books, on which modern opinions of his conception of science and his use of experiment have been based. In those books, presenting mature fruits of his science, Galileo's analyses are less causal and less experimental than is his analysis in the *Discourse.* The word *cause,* frequent in this early book, is less frequent in the later ones. It played little part in Galileo's mature presentation of scientific material, which he confined more and more to observational and mathematical statements. The single definition of cause sufficed for Galileo, who did not adopt traditional philosophical distinctions into material, formal, efficient, and final causes. Indeed, in his final *Two New Sciences* Galileo dismissed as fantasies all the speculations of philosophers about the cause of acceleration in free fall, saying that they did not merit the trouble of examination and that for the purposes of his science it sufficed to study uniformly accelerated motion, whatever might be its cause.

It is tempting to say that Galileo had banished causal inquiries from his science by the time he wrote that, and some have said he turned from the question *why* things occur as they do in nature to the question *how* they occur. There is nothing objectionable about such statements, which do sum up a very significant movement in physical science dating from the time of Galileo. With respect to him personally, however, they are too sweeping, as if he had lost all curiosity about what lay behind, or might lie behind, the observed regularities and the reasoned mathematical laws of nature. A more accurate statement would be that Galileo came to regard causal inquiries as speculative and as unnecessary in science, though of interest for their own sake and often useful as clues to discovery in science. Their place for him was not in science but in philosophy and theology as higher pursuits; as he said in his last book: "Such profound speculations belong to doctrines much higher than ours, and we must be content to remain the less worthy workmen who discover and extract from quarries that marble in which clever sculptors later cause marvelous figures to appear that lay hidden beneath those rough and formless exteriors."

The explicit definition of cause for use in scientific inquiries led gradually to the abandonment of causal goals in science. That was certainly not Galileo's intention when he introduced his definition into the dispute over floating bodies, nor was his definition noticed and adopted by later scientists. The process both for Galileo personally and for later science generally was an unfolding of terminological reforms and refinements, of which the beginning is discernible in the attempt by philosophers in 1611 to stifle scientific inquiry instead of pursuing it. Five years later they had sought and obtained the cooperation of theologians in another area of science, to Galileo's surprise and chagrin. He continued to respect philosophy and theology to the end of his life, as is illustrated by the foregoing quotation, but he continued to lose respect for contemporary philosophers and theologians who would not content themselves with remaining in their proper lofty pursuits but wished also to control matters that could be settled by "sensate experiences and necessary demonstrations." Such matters, in Galileo's view, belonged to science alone, and were all that properly belonged to science.

As time went on, others came to see things in that way, with perhaps little or no influence from Galileo. Causes remained under the exclusive jurisdiction of philosophers and theologians, while scientists focused their activities on laws of nature. In the middle of the eighteenth century David Hume presented to philosophers the consequences to philosophy of a definition of cause quite similar to Galileo's. Just as a new orientation of science had been associated with that at the outset, so new orientations of philosophy have become associated with it today, continuing the age-old interplay between cause, experiment, and science.

THE FIRST DAY

Interlocutors: *Sagredo, Simplicio, Salviati*
Scene: *Sagredo's palazzo in Venice*

Sagredo Welcome to Venice, most distinguished and learned Signor Professore! I am greatly honored that you have accepted my invitation and have troubled to come here from Padua on this occasion.

Simplicio Good morning, milord Sagredo, noble ambassador of our Most Serene Republic. It is many years since your presence graced our University when you were a student, and I was pleased that in the midst of your official duties you recall acquaintances of those days. Nothing could have kept me from coming when I learned that we shall have news today of my former colleague, the now celebrated Galileo Galilei, who I believe was once your teacher, years ago.

Sagredo He was indeed, and he still honors me with letters from time to time. Two weeks ago I heard from him that his friend and patron at Florence, the noble Signor Filippo Salviati, was coming to Venice, whence he will embark for Spain. Yesterday word came from Ferrara that he expects to arrive here today, so you shall have news also of the city in which your illustrious career as professor of philosophy began, continued now at Padua to the glory of my alma mater and of the great Republic it is my pride to serve. Tell me how things go there, for since my return from Crete I have had little news of the University.

Simplicio Things go quietly, since this is still the time of our long recess, allowing me time to write and polish my forthcoming book *On the Heavens.*

Sagredo In that case I must apologize for interrupting your profound studies and express redoubled gratitude for your presence here. I presume that in your book you will present a sound Aristotelian explanation of our friend's new discoveries in the skies, and particularly of those mountains on the moon that have perplexed philosophers, to say nothing of the wandering Medicean stars that accompany Jupiter.

Simplicio No, I shall say nothing of those. My purpose is to resolve doubts about Aristotle's meaning in disputed passages of his immortal *De caelo,* so that all problems of cosmology may be seen to have been solved—once Aristotle's text has been restored from errors of interpretation by certain expositors and commentators.

Sagredo But are not new problems raised by our friend's discoveries? I was absent from Venice when he made them, but surely you must have been among the first to whom he showed them here.

Simplicio Despite his repeated urgings, I would not look through that diabolical tube of his, which mightily displeased him.

Sagredo I see then why no new problems were created for you. But what were your reasons for refusing to use his telescope?

Simplicio Everyone knows that curved glass distorts vision. Why, then, should philosophers believe what cannot be seen except through two such glasses, one curved inward and the other outward? Even if I beheld new appearances, seemingly in the heavens, produced only by such means, then, knowing the heavens to be immutable, I should simply be obliged to explain the many illusions of optics. For that I have neither patience nor time for special study; nor could this have proper place in a book on the loftiest subject known.

Sagredo Even illusions appear to me to require explanation, and if I am not mistaken Aristotle did not disdain to mention them. If we are to use observation as the beginning of science and as its last court of appeal against speculative conclusions, it is necessary to know when vision may deceive us. And if I may speak frankly, learned professore, I believe that you use spectacles for reading and do not regard as illusions the words you then see through curved glass.

Simplicio Illusions do require explanation, but illusions in the heavens do not require explanation in a book on cosmology. As for my using spectacles, I say first that you would be surprised how many of the words I read through them are the merest and vainest illusions, which is why I find it necessary to publish the book I am now writing. And I add second that even ordinary spectacles give me a headache, so I have reason to believe that the two curved glasses of our friend's telescope would doubly do so, and I already have headaches enough in meeting the absurdities of commentators who have distorted the texts of Aristotle as badly as any curved glasses distort things seen. It is enough to see those correctly, and then nothing seen can remain to be explained.

Sagredo I envy you your certainty that the eyes of Aristotle perceived all that could ever be seen, for I, like our friend, am constantly assailed by new questions arising from observations, no less in terrestrial than in celestial phenomena. This began many years ago when he called my attention to the tides, a great motion seen every day here in Venice and yet never explained by Aristotle, who is even said to have so despaired of ever understanding them that he ended his life by plunging into the churning waters beneath the cliffs of the Negroponte.

Simplicio A childish fairy tale such as attends the biographies of all great men. Aristotle would have had little difficulty in accounting for the tides, had he judged them worthy of study in comparison with the profound contemplations of his First or Divine Philosophy, now called by some *Metaphysica,* which occupied his latest years. As to our friend's attempted explanation of tides by means of double motion of the earth, that is too fantastic to deserve a moment's notice, as I often told him when he attempted to interest me in it.

Sagredo My opinion is quite different, for I agree with him that motions should be attributed to other motions whenever possible and not, as by some in this matter of the tides, to an occult power of the moon's body over our seas. But there in that gondola, if I am not mistaken, comes our traveler. Let us return to the canal to greet him; see, the steersman is stopping here.

Salviati Hail, most noble Signor Giovanfrancesco Sagredo, if one of you indeed bears that distinguished name.

Sagredo I am him you seek. Welcome to our fair city of Venice and to my palazzo, which shall be yours as long as you will stay. Here, give me your hand and step ashore. Well, Signor Salviati, it is indeed a great pleasure for me to meet you in the flesh after having heard of you so often from our friend. Allow me to present to you his former colleague, first professor of philosophy at the incomparable University of Padua and the most celebrated living interpreter of Aristotle, who appropriately bears the very same name as Simplicio, Aristotle's most distinguished ancient commentator. Eminent professore, I have the honor of introducing to you the high-born Florentine Signor Filippo Salviati, who now supplants us in the affections of our departed friend, his countryman—though I hope not to our entire exclusion.

Simplicio I am delighted to meet His Excellency Signor Salviati, but I beg him to disregard the flattering and exaggerated courtesy of our host in his excessive praise of me. We were in fact heatedly disputing just now, which shows you how little the reputation he so graciously gives me protects me from his contradiction.

Salviati Rather, I shall join him in contradicting you, and on such short acquaintance too, against all rules of polite behavior; for I know that our esteemed host was guilty of no excess in his praise of you just now. Your name is celebrated throughout Italy, and beyond. Your splendid book *De Paedia* instructed me well in physics before I ever met your former colleague, our common friend, who has since told me many stories of your acuteness in defending the science of Aristotle. I am deeply honored and grateful to find you here to meet me, most learned professore, when I had planned a special visit to you at Padua to convey our friend's special greetings and compliments.

Simplicio Recalling as I do the vehemence of many of our disputes, I am both surprised and pleased to learn that he remembers me kindly. But now that we have exchanged compliments, let us all dispense with formal courtesies. It is true that I stand on ceremony at the university, as doubtless you both do at courts and in palaces, but here I should prefer us to speak together as we were wont to do with our old friend, who abhorred unnecessary formalities.

Sagredo It is indeed very gracious on your part, Simplicio, to offer us a freedom that I willingly accept—as I believe our distinguished visitor will also do, for I see by his face that he is pleased to hear your words. So, Salviati, come with us to seek a pleasant place in my garden where we may sit down and talk, for I think we shall spend many hours in conversation today; and, as we go, tell us of our old friend whom you left but a few days ago.

Salviati Our friend is well, except for distressing recurrences of that illness he acquired while here. The air of Florence brings them on at certain seasons, for which reason he comes frequently to live with me at my Villa delle Selve in the hills not far from Florence.

Sagredo Here is a warm but shaded place; let us sit, and refreshments will soon be brought. But does the grand duke permit our friend to absent himself from Florence and from the Tuscan court?

Salviati Indeed he does, though not without reluctance, for the wit and vivacity of our friend's converse much enlivens for Cosimo the dull and stilted round of courtly life. But he cares more about the health and comfort of his mathematician than about the strict performance of his new duties—which, as you may surmise, our friend nevertheless carries out punctiliously. Our sage young prince recognizes that more glory is likely to accrue to Florence from new researches and discoveries than from routine services, and he leaves our friend free to do as he wishes.

Sagredo I am happy to hear this, though not at all surprised. From his coronation four years ago Cosimo has shown, despite his youth, those qualities of true statesmanship and humanity that we in Venice admire above any conquests or displays of power. I rejoice also that you have given our friend a quiet place for study. And what researches now occupy him?

Salviati They are so numerous and varied as to defy description. At present he is engaged in perfecting his tables of motions of the Medicean planets for use in a certain project that he wishes kept secret for a time, while he is also composing replies to certain adversaries of his book on bodies in water, and that other on sunspots, both of which I presume you have seen and read.

Simplicio I did not read the one you mentioned first, as it treats of a subject hardly worth the time of a philosopher, besides which I have been told that it is full of absurd paradoxes. I did look at the work on sunspots but found it very disturbing, especially that final part endorsing Copernicus. I do wish our friend would stick to his mathematics and not meddle in the heavens with that infernal telescope of his, which is only going to get him into serious trouble sooner or later. I told him so before he left, but as usual he would not listen to reason, like a child with a new and dangerous toy.

Salviati Already there have been signs of such trouble; last year a Dominican father at Florence began saying that "Ipernicus" was contrary to the Bible. But the attacks against our friend's science have come from philosophers, not theologians, and would suppress any science that contradicts Aristotle. From what I have heard about you from him, you never retreated under attacks by ill-informed censors; no more will he, whatever their motivation.

Simplicio Doubtless you refer to an attempt made by the inquisitors here to compel me to alter a book of mine. It is true that I refused to do so, my job being to expound the opinions of Aristotle and not to determine whether any of them may be deemed heresies. Were I to attempt that and to fall into error, I might be in greater peril than I am. So I told them to go ahead and alter my book themselves, if they were so inclined, and to take responsibility in print for any theological statements without shifting that to me, who have no proper credentials for it. But surely that case is different. Aristotle wrote what he wrote, and as long as I correctly state his meanings, it is no concern of mine that others now judge any of them heretical.

Sagredo In my view, Simplicio, there is less difference than you think between your case and our friend's. His job is to expound nature, in which no heresies can possibly exist, nature being God's own obedient executrix. It is no task of his to determine what may offend philosophers among the things he sees and demonstrates. Nature shows some things that go against Aristotle, and our friend should not be asked to ignore or conceal them.

Simplicio How can you say this, when it is Aristotle alone who created natural philosophy? Clearly nothing that our friend can *see* in nature could possibly go against that. And when you say *demonstrates,* you must mean by pretended mathematical proofs, which Aristotle himself showed long ago to have nothing to do with physics.

Salviati I believe you allude to Aristotle's declaration that the mathematician and the natural philosopher (or physicist) may seem to concern themselves with the same thing but never deal with the same aspects of that thing. That proposition may be true without its following that mathematical demonstrations can have no place in physics. That seems to me a Peripatetic non sequitur rather than something found in Aristotle. One might as well say that grammar can have no place in poetry, since the grammarian and the poet both concern themselves with language, but never with the same aspects of language. And what if mathematics is the language of nature, as our friend conceives it to be?†

Simplicio I smile not in disrespect, Salviati, but because your way of speaking reminds me so strongly of our friend's that it is almost as if he sat here in your person. How happy this makes me that you have come, and that I have the good fortune to be present! For I always relished his sharp oppositions, however fiercely we quarreled, and I much miss them now that he is gone. Well, you have hit me hard, for nothing is so clear to me as the unwarranted extension by some Peripatetics of Aristotle's words to purposes for which he never intended them, and now I myself am accused of such a fault. To tell the truth, it never occurred to me until now to question the proposition that you challenge, it having been so long and so widely accepted. Your "language of nature" analogy now makes me stop to think, for though it does not appeal to me it does deserve attention. The debate whether the essence of nature is mathematical or substantial is as old as philosophy, while this middle ground comes as new to me, and I shall have to reflect on it before I reply. Will you be staying a few days, or do you leave for Spain at once?

Salviati This is my first visit to Venice, and I do not intend to leave without seeing more of its beauties than have been afforded me during my journey from the mainland. That alone would detain me, even had I not already planned to stay awhile, advised by our friend that nothing in Italy, or perhaps in the world, is quite like Venice. Already I find that not just sights, but the people here, and their customs, are utterly charming, to judge just from yourselves and from the witty boatmen who brought me here. In Florence we debate everything strenuously, gaining a reputation for pugnacity among those who do not know our customs. I came here rather expecting formality to the point of artificiality, and dreamy speculation in place of hard argument. But already I find myself so much at home, though enchanted by strange and lovely vistas, as not to feel any urge to leave at all.

Sagredo Why, then, you must make your home with me as long as you like, and I shall have the pleasure of conducting you anywhere you may desire in Venice and its islands. Far from this causing me any inconvenience, as you may fear, I have found that the presence of a visitor frees me from daily habits and takes me to parts of Venice I should more often visit. And you, Simplicio, shall join us here or on our tours as often as you wish, resuming your old controversies with our friend through his proxy here, if that suits him.

Salviati Thank you, Sagredo; I shall take you at your word and remain at least a few days. And though it will be small repayment for your kindness, I shall be happy to present to the best of my small ability the reflections of our friend. As I said, I have been privileged to spend much time in his company lately, especially at Le Selve, where he polished for publication his book on bodies in water and wrote his sunspot letters. Thus I am in a position to clarify, if you like, any doubtful points that may remain with you concerning those books.

Simplicio I too thank you, Sagredo, or rather I thank both of you for offering this engaging prospect during the next few days, in place of continued drudgery on the book that Sagredo and I were discussing when you arrived, Salviati.

Sagredo I shall take it not as small repayment, Salviati, but as payment with usurious interest for my little trouble if you shall reacquaint us with the thought of our friend by telling us the directions it has taken since his departure. And, to begin with the books you mention, I should like to know more about the circumstances of that dispute with philosophers at Florence out of which flowered his work on floating bodies. A few hints of this are found in that book, but tantalizingly meager ones—I suppose by reason of our friend's usual courtesty in not exposing to public blame the names of those with whom he disputed, associated with arguments that might now cause them embarrassment.

Simplicio That is an excellent suggestion, since, as I told you, I have not so much as glanced at that book. I have heard it it be full of paradoxes that, as a philosopher, I may be able to assist in resolving through the true principles of science.†

Sagredo I take it that by those words you mean the doctrines of Aristotle. Well, Simplicio, I think you may be surprised to learn that in this book our friend went out of his way to adopt one important principle from Aristotle even in preference to Archimedes, whom (as you know) he so greatly admires, so that any difficulties may lie in directions different from what you quite naturally assume. At the same time, however, he was not uncritical of Aristotle and also of his ancient rival Democritus. But perhaps Salviati will begin by telling us, from his conversations with our friend, what principles of science he now adopts and how they depart from Aristotle's.

Salviati Excuse me, Sagredo, but there is a certain ambiguity here in your use of the word *he.* If you mean our friend, I may be able to do this, tentatively at least, though I may not recall his exact words. If on the other hand you meant me, I can easily tell you my views, much altered since he returned to Florence, and now perhaps very like his own.

Simplicio Whichever *he* Sagredo may have meant, Salviati, let us hear both, in whatever mixture you think best. Your dissents will be as welcome to my ears as his conclusions, if not more so when they throw light on faults in such principles of his as may depart from Aristotle's exact words—for nothing has more injured true philosophy than careless paraphrasing, or strained interpretations imposed on texts of Aristotle's.

Sagredo Excuse me if I smile, Simplicio, and tell you that while our friend was still here he more than once said the same thing; and, much as he liked you, I do not think he excluded you from his complaint. So now, though the two of you often disagreed in natural philosophy, and not seldom with such violence as to alarm those who did not know you well, it may turn out—with Salviati as interpreter—that some of those altercations were formal and not material, depending when properly understood more on the sense or placement of a word than on substance. Let us hear, then, what Salviati has to say.

Salviati I shall begin by telling you that rather recently our friend has started using a phrase that sums up both his agreement with and his challenge to Aristotle, while at the same time it reveals his chief principle of science, in which (he says) nothing may contradict, and everything must have support from "sensate experiences and necessary demonstrations."† Whatever lacks either, he says, belongs to the humanities, poetry, philosophy, theology, or other realms beyond the reach of science thus restricted.

Simplicio The phrase is indeed interesting, and its implications for the most part are quite clear and self-explanatory. Thus his agreement with Aristotle is shown by insistence on necessary demonstrations, as taught in the *Posterior Analytics,* while what you call his challenge lies in an insistence on sensate experiences, or on mere facts, which Aristotle excludes from science properly understood, that being knowledge only of the reasoned fact. But surely you did not mean, as you seemed to say (perhaps by inadvertence), that our friend holds that what belongs to science does not belong to philosophy. Even in our fiercest debates he never went that far, nor do I think anyone would.

Sagredo In a way he did, Simplicio. Excuse me, Salviati, for you were about to speak, but first I want to remind Simplicio of one of those "fierce debates" here. Simplicio, in the fiercest of them all, which was over the new star of 1604, I distinctly remember that the basic issue was whether determination of its distance from the earth should be made by science or by philosophy. How then can this opinion come as anything new to you?

Simplicio You are very much mistaken, Sagredo. Our debate was whether in such matters the mathematician or the philosopher was to be trusted, not the scientist or the philosopher.† That merely gives two names to one thing, since natural science is nothing but a branch of philosophy. Mathematics, properly speaking, is not a science at all, but a mere preparation of the mind for philosophical reasoning. Some misguided philosophers tried, several centuries ago, to bring mathematics into physics, and one of those, a certain Englishman, was so fanatic about it that he is to this day known as "the Calculator," which is as far from being a true philosopher as anyone can get.

Sagredo These are but the winds of fashion, Simplicio, for not a century ago that Englishman's doctrines were studiously pursued by philosophers even at Padua. But let us not wander into a long digression. Accepting your correction, I withdraw my confusion of mathematics with science and say only that I believed our friend to hold a certain view at the time of that dispute. Now, since the question asked of Salviati concerns our friend's present view, let us hear that.

Salviati Dear me, gentlemen, I almost fear to say anything more at all, since if I do I must go back to where I was and again contradict Simplicio, which is hardly proper conduct for a guest who hopes to be well received.

Simplicio Oh, come, Salviati, you must know that contradiction is the daily fare of a philosopher, especially one in my position. At Padua it has long been the custom to appoint to the two highest chairs men who differ fundamentally on as many philosophical positions as possible, provided only that each shall represent a respected tradition. In my opinion that has been the source of Padua's strength, and I would not change it for anything. So I have a very thick skin and feel quite able to hold my ground, beneath which is the solid rock of Aristotle. Therefore speak frankly as our friend's proxy, for I begin to itch for that merciless combat of ideas that he formerly provided for me—to the alarm, as Sagredo says, of more tender spirits who feared we should come to blows.

Sagredo Simplicio is quite serious, Salviati; do not hold back. We Venetians are more Florentine than you think, to judge by your earlier remark. Perhaps you have been misled by Doni's amusing portrait of Italian cities contrasting their varying pleasures, in which those of Florence are fierce arguments in the red sunset whilst those of Venice are light, dreamy fantasies over the still waters of small canals. All guests in my house are free; none are invited who would give offense—or take it, either—in the flow of discourse.

Salviati Very well, then. Contrary to your assumption, Simplicio, sensate experiences constitute the part of science in which our friend *agrees* with Aristotle, who took them to outweigh any amount of reasoning, while the necessary demonstrations intended in our friend's phrase are mathematical proofs, not causal analyses of the middle term as in Aristotle's syllogistic approach. This constitutes our friend's *challenge* to Aristotle, which I hinted at earlier—that mathematics, rather than logic, is the language of nature and thereby of natural science.

Simplicio Now I see how much thought this new conception is going to require on my part. But you see how natural my assumption was, for, lacking such opponents since our friend left, I have lapsed into the supposition that every thoughtful person means by *science* just what Aristotle meant. But you, Salviati—surely you do not follow our friend in this attempted redefinition that would separate science from philosophy? I prefer to think you began with this in order that you might the sooner begin to reveal your own dissent.

Salviati Let me try to soften my reply by telling you that when our friend first came to Florence three years ago, I indeed shared the view you kindly ascribe to all thoughtful men. Perhaps I may add that in philosophy I was largely self-taught and remained content with wisdom I found in books by the most approved authors. When our friend arrived, I admired him as a mathematician and astronomer, thinking it strange that he had been appointed also as court philosopher. During the initial excitement over his new things in the heavens, he had no occasion to speak of his other researches into mechanics, and motion, and thereby physics and natural science. It was not until his return from Rome, where he had gone to exhibit the new discoveries, that he began to frequent the circle of philosophers who customarily met in my palazzo. There, one evening, the discussion turned to the difficult problem of condensation and rarefaction, out of which grew a debate on floating and sinking.† As this went on from week to week I saw how sensate experiences, introduced by a Peripatetic, became in the hands of our friend the basis of an entirely new extension of an old science, of which Aristotle had spoken hardly at all, but in which certain geometrical demonstrations led me to a firmer conviction than I had ever before known regarding physical questions. Thus, in the end, I abandoned the notion of science I had formerly held and came around to the view that seems so to disturb you.

Simplicio In so slight a matter as the floating of bodies, which as you say Aristotle hardly deigned to mention, no harm might follow. But did you not see, and does he not see, that as from small flaws great errors follow, all physics might be overturned by pursuing such a method, from attempts to base any part of it on so uncertain a mixed science as mere mechanics?

Sagredo If I may intervene, Simplicio, it seems to me that we should go at things the other way round. Instead of fearing to learn what happened in one small part of physics investigated by a method different from Aristotle's, lest other parts be affected or even the whole upset, we should make good use of our guest's stay here to learn what did happen, and only then judge whether the same method might confirm in new ways what is already known, or add to that, or even alter and improve parts of it.

Salviati Saying this, Sagredo, you put me in mind of another notion lately put forth by our friend—that investigations in science conducted independently of old principles may help us philosophize better. I do not know what you may think of this, Simplicio, but it seems to me that just as Aristotle first wrote his physics and then investigated its principles in his metaphysics, so in our day we would only be copying his example if we started from sensate experiences unknown to him, or at least unmentioned by him, and afterward attempt to philosophize about what, if anything, has been newly discovered.

Simplicio Aristotle was obliged to proceed as you say because there were no sciences before him, or none worthy of esteem, so that he had to found them all. Our situation is different, because all possible sciences have been so well established by Aristotle, and all their principles so clearly proved in his metaphysics, that it would be madness to question the latter and start all over. As he said, it is wrong to overthrow a principle unless you have a better one to put in its place.

Sagredo That seems to me like saying a country should have no men in its army save native-born veterans who have previously experienced battle. Within a generation such a country would have no army and would fall prey to any enemy lacking so laudable a principle, adopted to protect the inexperienced. For how will anyone find a new principle to put in place of some old one without first making new investigations that do not depend on that old one? Surely Aristotle contemplated such possibilities, or he would just have said that it is wrong to question any principle at all—and we know that he himself questioned principles of his predecessors, including even his own teacher, Plato.

Simplicio Very well; I am confident that no principle of Aristotle's can be successfully challenged by any method, and since we are discussing things among ourselves, out of earshot of young students who might be unsettled by our talk, let us hear, step by step, how this new method of our old friend led Salviati to reverse the sound view of science he had previously acquired from the best books. In that way I shall have opportunity to question and dispute any departure from Aristotle I detect, and may restore Salviati's feet to the path of truth when they stray from it. Have we time for this, Sagredo?

Sagredo Yes indeed, if Salviati will oblige us, since we may continue on other days if we wish. But I warn you, Salviati, that I too shall feel free to interrupt with questions of a different kind and purpose, namely, that of prying from you new information about the activities and thoughts of our friend. Is this agreeable to you?

Salviati Nothing will be more congenial to me than an attempt to impart to you what I have learned, representing to the best of my ability him who taught me. I take it that we shall speak only of things relating to his book on things that stay atop water or move in it, omitting that other, on sunspots.

Sagredo Let us do that, with the proviso that on some later occasion, perhaps on your return from Spain, we may meet again to explore the other book if this session comes out to our liking. And perhaps in years to come we may reconvene to discuss others he may publish, as I expect he will, on topics of still greater import.

Salviati That will give me a pretext for visiting your fair city again and again, as I have no doubt that such works are even now maturing in his mind, from hints he has given in discussions I have had with him. And now, to ensure that in our conversations nothing of importance escapes my memory, and to save the time that is often wasted in unplanned and rambling discourse, I have a proposal to make. You, Sagredo, surely have a copy of the book itself, and probably copies of both editions. If so, let the second and more complete edition be brought to us, from which we may by turns read the discourse from beginning to end, each of us being free to interrupt with questions or to add comments at any time.

Sagredo A servant has been sent to fetch the book. While we are waiting, Salviati, can you explain its rather curious title? Why did our friend not, in his usual simple style, just call it *Discourse on Floating and Sinking in Water*?

Salviati Ah, Sagredo, your perspicacity and your familiarity with his usual straightforward style have enabled you to see from the title alone that he was content neither with *floating* nor with *sinking* to describe the contents of his book. The reason for the latter is almost obvious, for rising in water is no less in need of explanation than sinking; perhaps more so. The discussions at my house that began the whole affair made this clear, for when philosophers asserted that ice floats because its flat shape prevents its piercing the resistance of water to division, our friend (who in any case denied the existence of such resistance) replied (as you will presently hear) that a flat piece of ice forced to the bottom will rise to the top of its own accord, despite such fancied resistance, though in that direction its own weight cooperates with flat shape to hold it still. I think it was that remark that first made me realize how badly all the philosophers of our circle had reasoned about the cause, which began my education in the new method of first examining all the observed effects.

Sagredo Here is the book, but I shall not let you have it until you finish answering my question.

Salviati What? Oh, yes, you mean about the word *floating.* That was replaced by the phrase *stay atop water* after the first night of our discussion, when we were really talking only about the ordinary kind of floating, long ago explained and demonstrated by Archimedes. In ordinary floating, or natural floating we might say, the body returns naturally to float if by violence it is pushed below the surface and then set free. No one had yet mentioned another kind of "floating" in which the body will not return, but will descend clear to the bottom when completely wetted after the gentlest push barely below the surface. Accordingly, our friend's first proposition to the philosophers was Archimedean, being worded with only natural floating in mind. Three days later another Peripatetic, who had not been present, promised to refute that proposition by ocular experimental demonstration. It turned out that this man, Lodovico delle Colombe, meant to argue not that the original proposition was valid only for natural floating, but that the existence of this second kind invalidated it entirely. There was some unpleasantness about this, but in the end our friend undertook to reduce both kinds of floating to a single principle. To round this out he included in his book still a third kind of "floating" that, like the second kind, had never before been analyzed, by Archimedes or anyone else. The phrase *stay atop water* is simple and perfectly general, while *floating* might be taken as meaning no more than what Archimedes had long ago explained. There is still another reason for the wording of this title, but I should prefer to speak of that only when we come to the phenomenon giving rise to it.

Simplicio I am glad to hear that our friend still retains something, at least, of what he learned in philosophy; that is, the necessity of making exact distinctions in science to avoid vulgar confusions. Since he had the poor taste to write in common Italian and not in philosophical Latin, however, I daresay his book will come to be known as just *Floating Bodies,* and foreigners who translate it so will lose this subtle meaning of his exact title.

Salviati I cannot resist replying that while you are right about his learning from philosophers the making of fine distinctions, he also learned that some use those mainly to slip between the fingers of an adversary who has caught them in error; and again, that when it suits their purposes, they turn to grand generalizations in order to hide some very subtle distinctions, made by nature herself, that they are unable to explain by their principles—though it befits the scientist to observe such distinctions with great care. But all these things will become clearer as we proceed, so let me begin reading:

[63] Because, Your Highness, I know that to let the present treatise be seen by the public when its subject is so different from what many expect (and from what I should already have put forth according to the intent I stated in my *Astronomical Message*)† might give rise to the notion that I had completely put aside my occupation with the new celestial observations, or that at any rate that I pursued those studies too slowly, I think it good to set forth reasons both for putting that off and for publishing this.

As to the former, a delay has been caused not merely by the latest discoveries of three-bodied Saturn and those changes of shape by Venus resembling the moon's, along with consequences that follow thereon, but also by the investigation of the times of rotation around Jupiter of each of the four Medicean planets, which I managed in April of the past year, 1611, while I was at Rome. There I finally ascertained that the first and nearest [satellite] to Jupiter passes through 8° and about 29′ of its circle in an hour, making its entire revolution in one natural day, 18 hours, and about one-half hour. The second goes in its orbit 4° and about 13′ per hour and makes an entire revolution in 3 days and about 13 and one-third hours. The third passes in one hour 2° and about 6′ of its circle and measures the whole in 7 days and about 4 hours. The [64] fourth and outermost passes in every hour 54′ and about one-half [minute] and completes its circle in 16 days and about 18 hours. But because the high speed of their returns requires very scrupulous precision in calculations of their positions for past and future times, especially if these times are several months or years, I am obliged to correct the tables of these movements with further and more exact observations than in the past and to bring them within the narrowest limits. For such precision the first observations do not suffice, not only because of the short periods of time [they covered], but because I had not found a way to measure with any instrument the distances between those planets, so that I made notes of their separations simply in terms of the diameter of Jupiter judged by eye, as we say. Those, although they did not involve errors of 1′ of arc, did not suffice for the determination of the exact sizes of the orbits of these stars. But now that I have found a way to take those measures with no error greater than a few seconds of arc, I shall continue observations until Jupiter goes behind the sun, which should suffice for complete knowledge of the movements and orbital sizes of these planets, and of some other consequences as well.

Salviati Here I might add that our friend has shown me his calculations and has explained how he managed to determine all the periods, starting from two remarkable positions in December 1610 and much aided by a "great conjunction" of all four stars on the night of 15 March 1611, shortly before he left for Rome, so that he was able there to start making tables of the motions in April. But at first he neglected a certain correction that, when made and then interpreted in the Copernican system, enabled him to detect and even to predict eclipses of these planets, as you doubtless saw in the appendix to his *Sunspot Letters.* Those eclipses are sometimes wonderfully different from any we see of sun or moon. But this has nothing to do with our subject, and I must hasten on—unless you wish to hear more.

Simplicio I believe that the less said about that Copernican folly, the better. I warned our friend before he left Padua about the dangers of such speculations.

Sagredo Though I, on the contrary, am dying to hear more about this, I agree that we must get on to those three kinds of floating.

Salviati There is just a bit more preface here:

I add to these things the observation of some dark spots that are seen in the sun's body, which, changing their position in that, offer a strong argument either that the sun revolves, or perhaps that other stars similar to Venus and Mercury circulate around the sun, sometimes invisible by reason of their small elongations, smaller even than that of Mercury, and visible only when they come between the sun and our eyes—or perhaps they indicate both things. Certainty on such matters [is sought and] must not be scorned or hidden.

Salviati A final paragraph was added in the second edition, our friend having been still undecided on the point when the book first went to the printer in March 1612:

Continued observations have finally assured me that such spots are materials contiguous to the sun's body, there continually produced in numbers and then dissolved, some in shorter and some in longer times, being carried around by rotation of the sun itself, which completes its period in about a lunar month—a great event, and even greater for its consequences.

Sagredo Would you tell us what consequences he had in mind? I did not notice any in the *Sunspot Letters* beyond this carrying around of sunspots.

Salviati I could indeed tell you, but that would take some time, and I thought we agreed not to digress into things unrelated to the matter of bodies in water. Moreover, this would offend against Simplicio's precept that the less said about any Copernican follies, the better.

Sagredo Very well; I catch your hint, and in any case I am already impatient to get on the main business at hand.

Salviati So was our author, who put in this extraneous material at the beginning only for the reasons you have already heard. Having finished the preface we have just read, he begins in earnest:

Now, as to the other matter, many reasons have moved me to write the present treatise, whose subject is the dispute I had some weeks ago with various men of letters in this city, after which (as

[65] Your Highness knows) many arguments took place. My main reason is your hint in praising the written word to me as a unique means of making known the true from the false, and the real from the apparent reason—much better than oral disputation in which one or the other, or frequently both parties to the debate, growing excessively heated and raising their voices too loudly, or not allowing themselves to understand, or carried far from the main issue by obstinacy against conceding anything to the other side, confuse themselves and their listeners with variety of propositions. Besides, it has seemed to me suitable that Your Highness should be informed by me, too, of everything that has happened in this contest, as you have already been told about it by others. And, since the doctrine I follow in the matter treated departs from that of Aristotle and from his principles, I considered that arguments against the authority of that great man, which among many persons renders suspect of falsity anything not coming from the Peripatetic school, may better be given in writing than by word of mouth; whence I resolved to write out this present *Discourse.* In it I hope to show that it is not by caprice, or by my having failed to read or understand Aristotle, that I sometimes depart from his opinions, but because reasons persuade me—and Aristotle himself taught me to find peace of mind in that of which I am persuaded by reason and not solely by the authority of the master. And it is a very true saying of Alcinous [*sic,* for Albinus] that philosophizing must be free. Nor, in my opinion, will it be entirely without utility to mankind to resolve our question dealing with whether the shape of solids affects their going or not going to the bottom in water, since it may be quite useful to know the truth of this matter when it becomes necessary to build bridges or other structures over water, something occurring mainly in affairs of great importance.

I therefore say that, finding myself last summer in conversation with men of learning . . .

Sagredo Let me interrupt, for we here have been curious to know who took part. Do you know?

Salviati Yes indeed, for the argument began at my house, in that circle of which I spoke, and I followed subsequent events with interest. Two professors of philosophy at the University of Pisa, Vincenzio di Grazia and Giorgio Coresio, began the debate in the way you will next hear. Professor di Grazia, a staunch Aristotelian, led the dispute, while Professor Coresio, being Greek and knowing the texts of both Aristotle and Plato by heart, took care that no misunderstandings should arise by reason of Latin translations and Italian paraphrases. Others there said little, while two men not present came to play main roles in the later arguments, as you will see. Now, to continue reading:

. . . it was said in the discussion that condensation is a property of cold, and the example of ice was given. I then said that I should have thought ice to be rarefied water, rather than condensed, since condensation gives rise to shrinkage in volume and increase of heaviness, but rarefaction to greater lightness and increase in bulk; now, when ice is formed, it is lighter than water, since it floats thereon.

Simplicio Rarefied water, indeed! This perverse habit by which our friend turns things topsy-turvy gave me no end of trouble when he was here, though I must confess that it often led me to see even more clearly than before how sound indeed are all of Aristotle's doctrines.

Salviati "More rare" or "less dense" does not imply "less solid," Simplicio, but merely "lighter." Iron is lighter than gold, but hardly less solid. To clarify comparisons of lightness in water, our author added here in the second edition:

What I say is evident, for since the medium takes away from the [66] total weight of a solid as much as the weight of an equal bulk of the medium (as Archimedes proves in the first book of his *On things that float on water*), when the volume of a solid is increased by distension the medium will deduct more from its total weight—and less when, by compression, it is condensed and made smaller in bulk.

Simplicio You say this was added in the second edition, Salviati, and yet it clearly supports Archimedes. Now, a friendly mathematician told me that the very first proposition proved in this book contradicts an assumption by Archimedes, that when a solid is submerged in water it displaces a volume of water equal to itself, or to its submerged part. But if our friend had found a flaw in Archimedes, why should he later return to support his doctrine? I should think he would instead have vaunted his superiority over the great Greek mathematician.

Salviati Your informant misunderstood. No flaw was found in Archimedes; rather, an assumption of his was further extended to embrace cases he did not consider, Archimedes having been concerned mainly with the floating of ships in oceans. But I should prefer to put off further discussion of this until we come to that proposition.

Simplicio By all means; let us get back to the story of the debate.

Salviati I see Sagredo nod assent, so I continue:

It was replied to me that this [floating of ice] comes not from greater lightness but from broad and flat shape, which, being unable to push through the resistance of the water, gives the reason ice does not sink. I then answered that any piece of ice, of any shape, floats on water, a clear sign that its being as broad and flat as you please plays no part in its floating. And I added that a very clear argument of this is our seeing a very flat piece of ice, pushed to the bottom in water, immediately rising to float, which would be quite impossible if it were heavier [than water] and its floating derived from a shape incapable of pushing through the resistance of the medium.

Hence I concluded that shape was in no way a cause of floating or sinking, but [the cause is] only greater or lesser heaviness in relation to water; and I said that all bodies heavier than water, no matter of what shape, would indifferently go to the bottom, while those lighter would indifferently float regardless of shape. Next I questioned whether those who believed the contrary to be true were not induced to think so by their seeing how greatly a variety of shapes may alter swiftness or slowness of motion; thus broad flat bodies descend in water much more slowly than those of more compact shape, when both are made of the same material. From that, one might be led to believe that shape could be so expanded as not only to retard but actually to prevent and remove any further motion, which I think false.

Concerning this conclusion, during many days that followed, much was said and various experiences were adduced, some of which Your Highness heard and saw; and in this *Discourse* you shall have all that was adduced against my assertion as well as what has since come to mind on this matter in confirmation of my conclusion. If this shall suffice to remove what I consider the false position, it seems to me that I shall not have wasted my time and effort; but, if that does not happen, I may nevertheless expect a different benefit, which is to arrive at knowledge of truth by hearing my [67] fallacies refuted and true demonstrations brought forth by those who hold the contrary opinion.

Simplicio That seems to me a laudable view, sincerely held by our friend. He often assured me in our most heated arguments, and showed by his actions on some occasions, that he contested points only to reveal error, whether the error was mine or his. After our dispute over the location of the new star he once said to me that though on that issue he had no doubt of his correctness in extending the mathematics of measurement to the heavens, I had convinced him that reliance on such rules must be examined separately in every case, as for instance whether the rule for fall of bodies near the earth's surface could properly be extended to bodies moving near its center, if any such motion existed; and he told me that in the book he would write some day on falling bodies he would take care to stress that caution.†

Salviati Since you bring up this matter of heated disputes, I should tell you that our friend often speaks affectionately of you as one from whom he learned much, particularly about limitations on scientific conclusions. The caliber of his adversaries in this present controversy was different, and he courteously suppressed in his published book some of their arguments that he had recorded in a little essay, meant only for the eyes of the grand duke, who had rebuked him for engaging in public contests outside the court. He showed me that essay while polishing the book for printing during his long stay at Le Selve, and, since I may have occasion to mention it again, I should say that it was written for a very different purpose than the book we are reading, being an apology to the grand duke and not a reasoned presentation of things that went beyond the sceince of Archimedes; nor was it written for the benefit of others studious of nature. But to go on:

And to proceed with the greatest ease and clarity I can, it seems to me necessary before anything else to explain what it is that is the true, intrinsic, and entire cause of the rising and floating of some solid bodies in water, or of sinking to the bottom; and this is the more necessary as I cannot be entirely satisfied with what Aristotle wrote on this subject.

Simplicio It is true that Aristotle hardly mentioned such matters, his lofty mind being occupied with true science, which leaves all such trifles to mere mechanics. I only wish that in descending to them our friend would take more care not to contradict Aristotle, as he was wont to do here, and as he says he will do again in this book.

Salviati You shall see that he nevertheless takes as his basis a ground advanced by Aristotle and ignored or rejected by Archimedes, using this then to extend to doctrine of the latter. Also, in a very important matter, he notes how his adversaries, though supposing that they were following Aristotle, actually took a position Aristotle had expressly rejected. Yet elsewhere he shows that Aristotle reasoned less well in a certain matter than his ancient adversary Democritus. In matters open to sensate experiences and necessary demonstrations, our friend holds that no authority need be accepted beyond that of our own eyes and our own powers of reason. But you will see this as we go on:

I say, therefore, that the cause by which some solid bodies descend to the bottom in water is the excess of their weight over the weight of water, while conversely, excess of weight of water over the weight of solid bodies is the cause that others do not descend, and even rise from the bottom and surmount the surface. This was subtly demonstrated by Archimedes in his books *On things that remain above water,* and it was taken up again by a very grave author—wrongly, however, if I am not mistaken, as I shall try to prove later in defense of Archimedes. With a different method and by other means I shall seek to reach the same conclusion [as Archimedes], reducing the causes of such effects to principles more intrinsic and immediate, in which are perceived also the causes of some admirable and almost incredible events, as that a very small quantity of water may raise up and sustain with its small weight a solid body that is a hundred or a thousand times heavier. And since demonstrative advance requires it, I shall define some terms and then explain some propositions from which, as from things true and noted, I may then serve my own purposes.

Salviati Here I interrupt the text for a moment to point out something that may otherwise pass unnoticed. By his term *demonstrative advance,* our friend means to characterize an important feature of his new science. Such sciences as mechanics and hydrostatics are ordinarily called "mixed sciences," following a hint of Aristotle in his *Questions of Mechanics,* where he says at the beginning that the problems are physical but the solutions mathematical. Our friend does not mean to contradict this "mixing," or any other classification of sciences that is useful. But he looks upon astronomy as a possible model for a new science of physics, and astronomy is not usually called a mixed science, being considered purely mathematical. Now, it was Ptolemy who founded this science . . .

Simplicio Pardon me; if there was a founder of mathematical astronomy, it was Eudoxus, not Ptolemy.

Salviati The astronomy of Eudoxus was indeed mathematical, but in a different way from that of Ptolemy, for Eudoxus stopped with hypothesis and did not go on to prediction by calculation.† Our friend envisions a science of physics that, like the astronomy of Ptolemy, may successfully predict by calculation. Such sciences he calls *demonstrative,* and they advance by deducing, from hypotheses drawn from phenomena, what measurable consequences should follow. Such deductions constitute *demonstrative advance* in his sense. Whenever new and measurable consequences emerge from this, it becomes possible to test anew the tenability of hypotheses, and if necessary to correct them or abandon them for others that better meet such tests. Let us now return to the text:

Accordingly I shall call *equal in specific weight* those materials of which equal bulks are equal in weight. For example, if two balls, one of wax and the other of some wood, being equal in volume are also equal in weight, we shall say that that kind of wood, and wax, are equal in specific weight.

But I shall call *equal in absolute weight* two solids that weigh the same, however unequal in bulk. For example I shall say that a bulk of lead and another of wood, each weighing ten pounds, are of equal absolute weight, even though the volume of the wood is much greater than that of the lead—*and consequently* [*wood*] *is of lesser specific weight.*

Simplicio This term *specific weight* is ill chosen, though I approve of making a distinction between two ways people commonly say that one thing is heavier than another, saying at one time that lead is heavier than wood and another time that a particular wooden ball is heavier than some leaden one. But this has nothing to do with being "heavier in species," as if weight were sufficient to determine a species of material. If adopted, *specific weight* will even further confuse the difficult concept of *species,* so essential to philosophers.

Sagredo In this case our friend did not invent the term, which was introduced in Latin centuries ago in a book ascribed wrongly to Archimedes himself. A great mathematician of Brescia, Niccolò Tartaglia, who long resided and taught privately in our fair city of Venice, gave the term currency in Italian, so that it is well known among writers of this science, if not among philosophers. At Florence also someone appears to have similarly objected to the use of the word *moment* in this book, so in the second edition our friend added a further explanation.† It seems to me high time for philosophers to recognize and accept current names for things that were not considered by Aristotle.

Salviati You are right about *moment,* to the definition of which we are now coming. Our friend told me that the objection you mentioned was raised by the newly appointed professor of philosophy at Pisa, Flaminio Papazzoni, who entered the debate a month or two later, reluctantly, and only to defend Aristotle before the court. He then wrote a small book in reply to this one, concealing his identity in order to avoid the appearance of ingratitude to our friend, who had secured his appointment at the University of Pisa.

Back to our reading:

I shall call a material *specifically heavier* than another, of which an [68] equal bulk weighs more than a given bulk of the other. Thus I shall say that lead is specifically heavier than tin because, taking equal volumes of each, that of lead weighs more. But I shall call a body *absolutely heavier* than another if it weighs more than the latter, without regard to volume; and thus a great wooden beam is said to be more heavy absolutely than a little bulk of lead, though lead is specifically heavier than wood. And likewise are to be understood *less heavy specifically* and *less heavy absolutely.*

Those terms defined, I take from the science of mechanics two principles. The first of these is that weights equal absolutely, moved with equal speeds, are of equal forces and moments in their operation.

Salviati It is here that in the second edition our friend, having previously assumed that everyone would understand this, added, in reply to the objection of Professor Papazzoni:

Moment, among mechanics, means that force, that power, that efficacy, with which the mover moves and the moved resists, which force depends not simply on weight, but on speed of motion [and] on different inclinations of the spaces over which motion is made—for a heavy body descending makes greater impetus in a steeper space than in one less steep. And in short, whatever be the cause of such force, it always keeps this name of *moment.* Nor did it appear to me that this sense must come as new in our language, for if I am not mistaken we often say "This is indeed a serious matter, but that other is of little moment," and "Let us consider light things and leave out those that are of moment"—metaphors that I think are taken from mechanics.

So, for example, two weights absolutely equal in heaviness, placed on a balance of equal arms, remain in equilibrium, and one does not go down and raise the other, because equality of distances of both from the center on which the balance is supported and around which it moves would make those weights pass equal spaces in the same time when the balance is moved. Hence there is no reason at all that one such weight, any more than the other, should go down; and therefore they make equilibrium, and their moments of force remain similar and equal.

Sagredo Our friend once remarked to me that such must have been the consideration that induced Archimedes to put the equilibrium of such weights as his first postulate. Some say he did this as the result of repeated testing by experiment; but then he would have required some other definition of *equal weights,* and as a mathematician, among the greatest that have ever lived, Archimedes could not have failed to note this. Therefore, since he gave no other definition, and since no means existed for knowing which weights were equal save their resting level in a balance of equal arms, either this postulate was also his definition of *equal weights,* or he was guilty of a *petitio principii*—a glaring fault that no one would ascribe to Archimedes who had read his admirable writings.

Simplicio But that is only to save Archimedes from one fault in logic and charge him with another, perhaps worse one; namely, failure to define his terms.

Salviati Let me reply, Sagredo, showing Simplicio why our friend thinks differently. He admires the efficiency of stating a single postulate, as against first saying "I define weight as 'equal' in two bodies that remain at rest when placed in the pans of a balance of equal arms" and then postulating that in fact this definition applies to actual bodies observed in practice. He says that definitions either introduce new terms or restrict the use of familiar terms, but that they assert nothing about the existence of things defined, our friend following Aristotle in this. That is why Euclid, after he had defined a square, took care to show, before putting the term to use, that a square can be constructed. But in mechanics, says our friend, and even more in physics, we speak mainly of things actually existing in nature as asserted by general usage of their common names. We are not obliged, and indeed we are powerless, to construct the things named; those are presented to us by nature, which in this respect does for us what Euclid was obliged to do for squares in pure geometry. Hence, in our friend's opinion (with which I go along), we are entitled to take from nature and from our language some things, and from mathematics other things, as already given and to put these together in science without vain repetition of words. He thinks that science has been impeded in the past by two delusions, one of which arises from words offered in formally correct proofs after having been deprived of content as applied to things around us, deluding us into thinking that nature must obey our arguments†—and indeed he has heard philosophers vainly try to argue the Medicean planets out of the sky. The other delusion is that necessary demonstrations (by which he means mathematical proofs) may be left out of account in physics because mathematicians do not deal with material things, while physicists do; for it may sometimes *happen* that mathematical rules exist to which nature *does* conform, and then it behooves physicists to beware of the consequences of insisting on principles inconsistent with those rules.

Simplicio Many things that could be said occur to me, but let us go on.

Salviati Very well. Next he says:

The second principle is that the moment and the power of heaviness is increased by speed of motion, so that absolutely equal weights conjoined with unequal speeds are of unequal power, moment, and force, in the ratio of one speed to the other. We have a very suitable example of this in the [steelyard or] balance of un- [69] equal arms, on which absolutely equal weights being placed, they do not press and exert force equally, but that which is at the greater distance from the center (about which the balance moves) goes down, raising the other; and the motion of that which rises is slow, while the other is swift. Now, the power and force conferred by [greater] speed of motion on the moveable† receiving it is such that this can exactly compensate the additional weight in the other, slower, moveable. Thus if one arm of the balance be ten times as long as the other, so that in motion of the balance around its center that end goes through ten times the space of the other end, a weight placed at the greater distance can sustain and equilibrate another ten times as heavy, absolutely, as itself; and this is because when the steelyard moves, the smaller weight is moved ten times as fast as the greater. It must, however, be always understood that the movements are to be made at the same slope; that is, that if one moveable moves vertically, the other likewise makes its motion vertically, and if the motion of one must be made horizontally, the other also shall be made in the same plane, and in general both [are considered] at similar tilts. This equalization between weights and speeds is found in all mechanical instruments and was considered as a principle by Aristotle in his *Questions of Mechanics*.† Hence we also may take it as a most true assumption that absolutely unequal weights mutually counterweigh and are rendered of equal momenta whenever their weights correspond in inverse ratio to their speeds of motion, so that however much less heavy one is than the other, by that much it is constituted to move more swiftly than that one.

These things explained, we may at once begin to investigate which are the bodies that can be totally submerged in water and go to the bottom, and which ones necessarily float on top, so that if driven forcibly below they return to float with part of their bulk raised above the surface of the water. This we shall do by looking at the reciprocal action of these solids and the water, which action follows on immersion; and it is that in submersion of a solid drawn downward by its own heaviness, water comes to be driven out from the place into which the solid successively enters, and the displaced [70] water is lifted and rises above its first level, which lifting is resisted by its nature as a heavy body. And since the solid in descending becomes more and more submerged, a greater and greater amount of water being raised until the whole solid is plunged in, one must compare the moments of resistance of the water (against its being lifted) with the moments of the pressing heaviness of the solid; and if the moments of resistance of the water before total immersion shall equal the moments of the solid, then doubtless equilibrium will be reached and the solid will submerge no more. But if the moment of the solid shall always exceed the moments with which the displaced water successively resists, the solid will not merely entirely submerge in the water but will descend clear to the bottom. Finally, if at the very point of total submersion equalization shall be reached between the moments of the solid pressing down and [those of] the resisting water, then rest will supervene, and this solid will be able to remain at any place in the water.

So far, the need to compare the weights of water and of solids has been made manifest, and at first glance such comparison might seem sufficient for determination and conclusion as to which bodies float and which go to the bottom. We might say those float that are less in specific weight than water, and those sink that are greater in specific weight, inasmuch as it appears that a solid in submerging will raise as great a volume of water as that part of its own body that is submerged, whence it would be impossible for a body of less specific weight than water to submerge itself entirely, it being unable to lift a greater weight than its own, as would be a volume of water equal to it in bulk. And likewise it would appear necessary that a solid of greater weight [than water] would go to the bottom, as having more than enough power to raise a volume of water equal to its own, but of less weight than its own. Yet this business proceeds otherwise, and though the conclusions are true, the causes thus assigned are defective, nor is it true that the submerging solid thrusts aside and raises a volume of water equal to its own submerged bulk.

Simplicio Please stop a moment, Salviati, for I have a question and an objection. My question is what our friend means by saying that the causes assigned are defective, though the conclusions are true.

Salviati The conclusions to which he refers are those concerning bodies that float, those that sink, and those that would remain still at any place in water, totally submerged. Those are indeed respectively bodies of lesser, of greater, and of equal specific weight relative to water. The conclusions are true, were known to Archimedes, and will presently be demonstrated by our friend. But the causes alleged just now, that no body can lift a greater weight than its own, and that every body of greater specific weight than water, in any situation, can raise a volume of water equal to its own volume, are defective; for though they can be said to be true in a sense, or for the most part, yet instances exist in which they do not apply. These come about when one weight is a solid and the other a liquid, because no part of a solid can be lifted without raising all parts, whereas part of a liquid may be raised separately from the rest, yet without breaking the continuity of the whole. Such cases were not investigated by Archimedes, who also did not employ the principles our friend adopted or the causes alleged above, for reasons you will later understand. Note that our friend did not say the causes are false, but only that they are defective; that is, when amended they will serve for every case. Now, what was your objection?

Simplicio The objection is that, little as I know of geometry, it is evident even to me that one cannot seriously believe that any solid can occupy the space left for it by material smaller than itself. I fear our friend will have few followers in this new doctrine of his, or rather this new paradox, and will be laughed to shame by sound philosophers.

Salviati Your customary prudence has deserted you, Simplicio; first hear our friend out, and then judge his paradox. For a paradox there is here (though not the one you think), if by *paradoxes* we mean not only falsehoods that are made to appear true, but also truths that appear to be beyond human understanding. Here is what our friend says next:

> Rather, the water raised is always *less* than the submerged portion of the solid, and proportionately less as the vessel containing the water is narrower, so that no contradiction exists in the entire submersion of some solid under water without its raising more water in volume than would make up the tenth or twentieth part of [71] its own bulk. And on the other hand a very small amount of water may be able to lift up a very large solid bulk, though that solid should absolutely weigh one hundred or more times as much as that water, provided only that the solid be of material of lesser specific weight than water. Thus a huge beam weighing, say, 1,000 pounds, may be raised and sustained by water that does not weigh 50 pounds, which will occur when the moment of the water is compensated by its speed of motion.

Simplicio Again I protest, for we were talking about floating, which takes place in still water that has no speed, or any motion even.

Salviati And again I ask that you hear him out, for in due course you shall learn of the speed he means, which, like the speed intended by Aristotle in *Questions of Mechanics*, is not absolute and actual, but relative and potential speed, similar to that of either end of the steelyard when in equilibrium, as already seen. But let us read on, and perhaps the necessary patience will be inspired in you by his next words, which are then followed by a long demonstration omitted from the first edition.

Now, since such things offered abstractly as above present some difficulty to their understanding, it is good that we shall illustrate them by particular examples, and for ease of demonstration let us assume that the vessels into which the water is to be poured are enclosed on all sides within walls perpendicular to the horizontal plane, and the solids placed in those vessels to be right cylinders or right prisms.

This much explained and assumed, I shall demonstrate the truth of what I have suggested, in the shape of the following theorem:

The volume of water that is raised by submerging a solid [right] prism or cylinder, or that is lowered by taking that out, is less than the volume of the submerged or extracted solid, having to it the same ratio that the surface of the water surrounding the solid has to that same surrounding surface plus the base [area] of the solid.

Sagredo Excuse me, Salviati. Before we proceed to the demonstration of this theorem I think we should explain to Simplicio, who has perhaps not read the books of Archimedes, how and why our friend's assumptions differ from those of Archimedes and yet in no way contradict them.

Salviati You are right, Sagredo. Know, then, Simplicio, that Archimedes appears to have been concerned with the floating of ships in oceans and the conditions of their stability against capsizing. We can judge his purpose from his second book on bodies in water, for which the first book is a preparation, and we know from Plutarch's *Life of Marcellus* that Archimedes was no less, nay, if anything even more, renowned in his time for practical applications of his knowledge than for his prowess as a mathematician.

Plutarch said also that Archimedes disdained to write of mere mechanical applications, as plebian and unworthy of a philosopher; yet his only evidence for this (Plutarch having written long after ward) must have been that no such books survived, from which he deduced that none were written. But Cicero took back to Rome a certain mechanical model of the heavenly spheres made by Archimedes, who is known to have composed a book on its construction. This throws grave doubt on that snobbery ascribed to him by Plutarch, which was more probably the product of Plutarch's era than of that marvelous period of invention in which Archimedes flourished. At any rate King Hiero of Syracuse repeatedly praised the mechanical inventions and discoveries of Archimedes and, far from regarding them as plebian, associated them with the noble art of war. Plutarch himself well knew that they saved Syracuse more than once from conquest by the Romans, yet he permitted himself a snobbish comment discrediting useful science.

To get back to the subject, Archimedes wrote of floating in open seas, not in restricted vessels with vertical sides as does our friend. There is no doubt of this, since Archimedes proved the water surface to be spherical, and his diagrams showed water extending to the center of the earth, as even oceans do not do. That is why Archimedes had no need to bring in the ratio of which our friend writes in this theorem, and in science it is superfluous to bring in anything not needed; nay, worse, it is meaningless. The rise of water level of an ocean when a ship is launched exists only philosophically or abstractly, not measurably. Just as Ptolemy made certain calculations without regard for the sun's parallax, saying so and remarking that for his immediate purpose that would needlessly complicate the problem, so Archimedes omitted a change of water level that, as you will see, became a special object of our friend's scientific inquiry. Thus you might raise the water level two feet in a tall, narrow cylinder by thrusting in your walking stick, but you are in no danger of getting your feet wet by thrusting the same stick into the Grand Canal while standing at the edge of the Piazza San Marco at high tide. The reason for that is no less interesting than the reason for the stability of sailing ships, and leads to knowledge no less useful, but it did not bear on that problem.

Here I may add that our friend, in the essay he first wrote for the grand duke, employed only the simpler assumption, knowing that Cosimo would find it easier to follow without any demonstration of the present theorem. That assumption sufficed for the balance of his essay. But in publishing to the world our friend prefers to present only new things, so far as possible. Hence he began with a more general consideration of displacement than that of Archimedes, whose assumption then emerges as a particular case; that is, when the ratio of which our friend speaks is (or very nearly equals) the ratio of equality. This it does when an ocean is considered, though a fleet of ships should be launched in it, and also for a washtub in which a nail is placed. But if we place a wooden beam in a watering trough, the questioned ratio is very small, and from that fact arises a remarkable and useful pehnomenon to be discussed in due course. To use a phrase of Aristotle's, the assumption of Archimedes holds "for the most part."

Simplicio Thank you; your last remark has awakened my mind, for Aristotle used that phrase mainly to avoid the necessity of discussing uncommon exceptions not worthy of detailed investigation by his great mind. You mean that when the water level is altered measurably, the volume of water displaced is measurably less than the submerged solid, the very rise of water creating, as it were, additional space as measured from the final level. I take it that our friend's science is concerned only with measurable differences, however, for in true philosophy there would always be a difference, whether measurable or not.

Salviati Granted; and that is why our friend did not impute any error to Archimedes or trouble with this particular in writing his essay for Cosimo, for in neither case did the simpler Archimedean assumption affect the analysis involved.

Simplicio But now I am puzzled that our friend troubled with it at all, having in hand the assumption of Archimedes that serves in general and for the most part—especially when this adjustment of our friend's is so minute.

Sagredo Here, in my opinion, the puzzle stems from your words *general* and *minute.* The word *general* had two meanings, in one of which it has the sense of *by and large,* and in the other, *universally applicable.* We have now spoken of the first, since that is just another way of saying *for the most part,* which usually leaves the mathematician with a certain yearning to complete the investigation. As to the meaning *universally applicable,* the more our friend directly investigates nature, the less inclined he has become to assume that universal assumptions ever do apply when physical problems can be stated in termini or with mathematical precision. To some extent Aristotle was on the same side when he questioned the suitability of mathematical methods in physical studies. The difference is that our friend, seeing Aristotle brush these methods aside from physics, conceived that the discarded crumbs might possibly be enough for a meal. In my opinion he has already found them sufficient for several great feasts, with reason to expect more, and greater ones, in years to come.

Next, as to the word *minute,* I believe you think of this as meaning *trivial.* But until we investigate minute things we have no way of knowing whether or not they are trivial. If you will excuse my saying so, most people would consider "minute" the differneces between Alexander of Aphrodisias and Averroës in interpreting Aristotle, though to you those differences are anything but trivial. The possible importance of seeming minutiae must be judged not by outsiders, but by experts who have investigated them in any given field.

Salviati No better illustration of Sagredo's remark can be found than a later mechanical consequence our friend has reached from his minute adjustment to the equal-volume assumption.
But let me proceed; as I said, this is in the second edition only:

Let the vessel be ABCD, and in it let water stand at the level EFG before the solid prism HIK is immersed in it, while, after it is immersed, let the water be raised to the level LM. Then the solid HIK will be all under water, and the volume of water raised will be LG, which is less than the volume of the submerged solid (that is, of HIK), for it is equal only to the part EIK that lies below the original level, EFG. This is manifest, for if the solid HIK is drawn out, the water LG will return into the place occupied by the volume EIK where that was contained before submersion of the prism. The volume LG being equal to the volume EK, we may add to both the volume EN; and the whole volume, EM, composed of the part EN of the prism and the water NF, will equal the whole solid HIK. Hence volume LG will have to EM the same ratio as to the volume HIK. But volume LG has to volume EM the same ratio that surface LM has to surface MH, from which it is evident that the volume of water raised, LG, as the same ratio to the volume of the submerged solid HIK that the surface LM (which is that of the water surrounding the solid) has to the whole surface HM, composed of the said surrounding water [surface] and the base [area] of the prism HN.

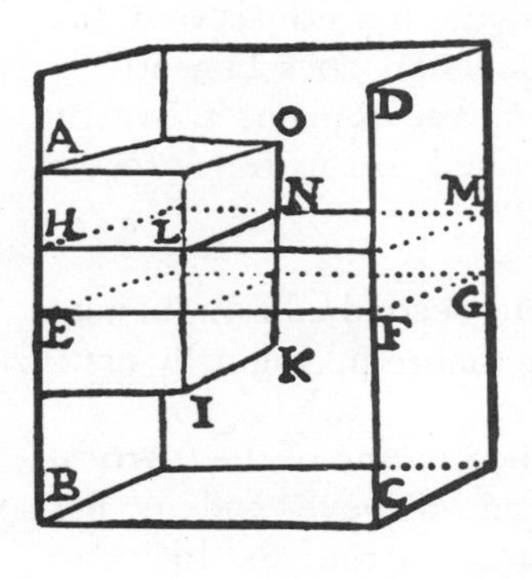

[72]

But if we suppose HM to be the original level of water, and the already submerged prism HIK to be withdrawn and lifted to EAO, while the water is lowered from its original level HLM down to EFG, it is manifest (the prism EAO being identical with HIK) that the upper part HO will equal EIK below, subtracting the common part EN. Consequently the volume of water LG will equal the volume HO and will be less than the solid when outside the water, that being the whole prism EAO; and likewise this lowered volume of water, LG, has the same ratio to it as the surrounding surface of water LM has to that same surrounding surface with the addition of the base of the prism, AO, the demonstration of this being the same as in the previous case.

From this it is seen that the volume of water raised by immersion of a solid, or lowered by its withdrawal, is not equal to the whole volume of the solid submerged or removed, but equals only that part of it which in the process of submersion goes beneath the original volume of water, and when removed remains above the original level, which was to be demonstrated. Let us now proceed to the rest.

Salviati As I said, this demonstration is only in the second edition, being placed immediately before the following theorem, originally first:

And first we shall demonstrate that when in one of the [vertical] vessels previously assumed, of any size and however wide or narrow, a [vertical] prism or cylinder is placed surrounded by water, then if we lift that solid vertically the surrounding water will be lowered, and the lowered water will have the same ratio to the raised solid as one base of the prism [or cylinder] has to the surface of the surrounding water.

Let there be such a vessel containing the prism ACDB and otherwise filled with water to the level EA. Let the solid AD be raised to GM; the water will fall from EA to NO. I say that descent of the water, measured from the line AO, has to the ascent of the prism, measured from the line GA, the same ratio that the base GH of the solid has to the surface of water NO.

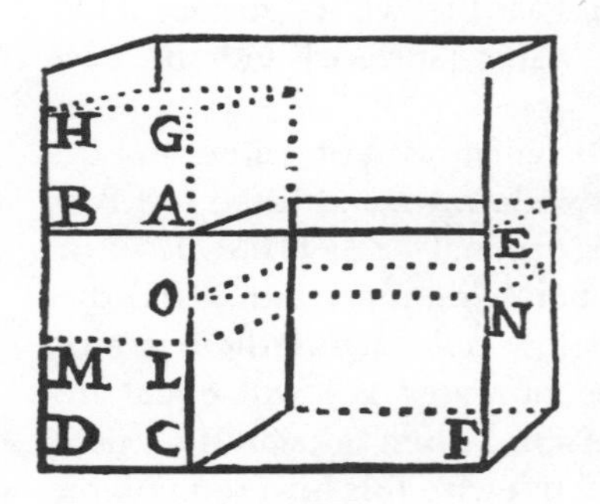

This is evident, for the volume of the solid GABH raised above the original level EAB equals the volume ENOA of water lowered, whence we have two equal prisms ENOA and GABH; but of equal prisms the bases [73] are inversely proportional to the heights, and therefore as height OA is to height AG, so is the surface or base GH to the water surface NO.

When, therefore, a column (for example) is placed upright in a very large pond, and again in a well capable of holding little more than the volume of that column, which is then raised out of the water, the surrounding water will go down in proportion as the column is lifted; and the lowering of water will have the same ratio to the distance through which the column is lifted as the area [*larghezza,* "largeness"] of the column has to the excess of area of the pond or the well over and above the area of the column. So if the well is but one-eighth wider than the column, and the area of the pond is twenty-five times that same area [of the column], then in raising the column one *braccio,* the water in the well would drop seven *braccia* while that of the pond would go down but one twenty-fourth of a *braccio* [or nearly fourteen feet against one inch].

Salviati Behold, Simplicio, how from the seeming minutiae of this purely geometrical investigation there has emerged a difference of effect that is anything but trivial, and one that we have all seen in various circumstances without thinking about it. But that is not all; you will next see how from its analysis our friend has arrived at what he calls the true cause of a very remarkable phenomenon, in which a lesser weight directly raises a greater weight—something Aristotle supposed possible only by use of the ingenious mechanical instruments devised by builders. But in this case it is effected by nature herself, without any miracle such as might seem to have been required by the writer of *Questions of Mechanics.* Our friend now says:

This demonstrated, it will not be hard to understand, through its true cause, how a right prism or cylinder of material having less specific weight than water, when completely surrounded and covered by water, will not remain so but will be raised, even though the surrounding water may be little in amount, and in absolute weight as much less than the weight of the prism as you please.

Let there be in the vessel CDBF the prism AEFB of less specific weight than water, and let water then be poured in to the full height of the prism. I say that the prism, being set free, will rise, driven by the surrounding water CDEA. For the water CE being of greater specific weight than the solid AF, the absolute weight of the water CE must have a greater ratio to the absolute weight of the prism AF than the volume CE has to the volume AF, inasmuch as those two volumes have the same ratio as the two absolute weights when [and only when] they are volumes having the same specific weight. But volume CE has the same ratio to volume AF as the surface of water CA has to the surface or base of prism AB, which is the same ratio as that which the ascent of the prism when raised has to the [attendant] descent of the surrounding water CE. Therefore the absolute weight of water CE has a greater ratio to the absolute weight of prism AF than the rise of prism AF has to the drop of the water CE. Consequently the moment compounded of the absolute weight of water CE and its speed of dropping, when it exerts force by pressing down and driving out or extruding the solid AF, exceeds the moment compounded of the absolute weight of the prism AF and its slowness of rising, that being the moment by which it opposes extrusion and the force exerted against it by the moment of the water. Hence the prism will be raised.

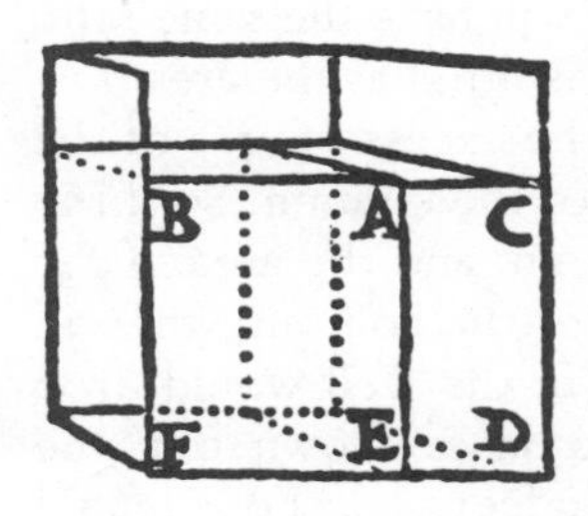

[74]

Sagredo I note that in the first edition our author proceeded at once to make use of this idea of *moment* after naming it, without the interposition of a purely geometrical theorem as in the second edition. Here, then, is the place to speak of that objection raised earlier by Simplicio against our author's appeal to "speeds" when speaking only of floating, which takes place in still water and without motion, let alone speed.

If I am not mistaken, Simplicio, the word *moment* first began to be used by mathematicians writing of centers of gravity and hence of conditions of equilibrium, in which by definition there is no motion. But here, in saying "no motion", there is perhaps something a little different from mere privation in the sense in which you philosophers like to employ that term. We expect a heavy body to move downward unless it is at rest in its natural place, which is as near as it can get to the common center of heavy things, or center of the earth. Since heavy bodies are endowed with this tendency to move down, and to do so with some definite speed whenever they are free to move, mathematicians have found it convenient to treat this potential speed as if it were actual, but only through some imperceptible distance. In that way they have been able to arrive at mathematical rules that are then found to apply consistently, as nearly as we can measure their implications for actual bodies. But up to now, except in the *Questions of Mechanics,* physicists have been chary of appealing to any speed without *actual* motion, because of logical inhibitions imposed by philosophers. That is why our friend, admiring the boldness of Aristotle in the *Questions of Mechanics,* took pains to credit him with the method adopted here in preference even to that of Archimedes. In contrast, the Flemish engineer Simon Steven,† who thus far has published more than anyone else of our age to advance the sciences of mechanics and hydrostatics, to say nothing of mathematics itself, objects most strongly against reasoning from speeds where no motion exists, he denying even potential speeds through virtual distances in considering phenomena of equilibrium. So you see that to press your objection against our author would be in a way to take sides against Aristotle, which I believe you will not wish to do.

Simplicio Indeed not, and I see that in this case, as you tried to warn me, I was too anxious to protect Aristotle against fancied attacks before hearing out our friend's fuller explanations. I suppose my past experience of his questioning everything, even Aristotle, has made me resist whatever he proposes that comes as new to me.

Salviati As he said, philosophizing must be and remain free. If you are confident that everything true can be reconciled with Aristotle, whether or not explicitly said by him, then you above all should have the courage of that conviction. What our friend attacks is neither Aristotle, nor any appeal to new truths in support of Aristotle's principles, but the tiresome appeal to Aristotle's principles in support of truths that can stand on their own feet, grounded not on any man's principles but on sensate experience and necessary demonstration. To continue:

Now let us proceed to demonstrate more particularly the extent to which solids of less [specific] weight than water will be raised; that is, what part of them will be above the surface of the water. First, however, it is necessary to prove the following lemma.

The absolute weights of solids have the compounded ratios of their specific weights and their volumes.

Let there be two solids, A and B; I say that the absolute weight of A has to the absolute weight of B the ratio compounded from that of the specific weight of A to the specific weight of B, and that of the volume of A to the volume of B.

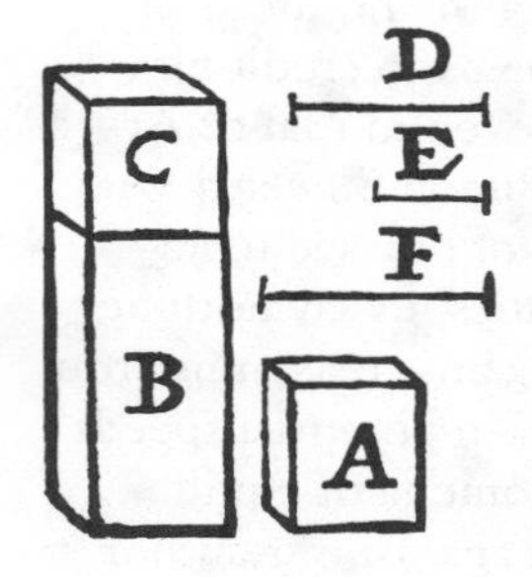

Let line D have to line E the same ratio that the specific weight of A has to the specific weight of B, while line E shall be to line F as the volume of A is to the volume of B. Obviously the ratio of D to F is compounded from the ratios of D to E and of E to F; therefore it must be demonstrated that as D is to F, so is the absolute weight of A to the absolute weight of B.

Take the solid C, equal in volume to A and having the same specific weight as solid B. Since A and C are equal in volume, the absolute weight of A will have the same ratio to the absolute weight of C that the specific weight of A has to the specific weight of C—or of B, which is the same in specific weight [as C]. But this is as line D is to line E. And since C and B are of the same specific weight, then as is the absolute weight of C to the absolute weight of B, so will be the volume of C (or the volume A) to the volume B, and that is as the line E to the line F. Therefore, as the absolute weight of A is to the absolute weight of C, so is line D to line E, while as the absolute weight of C is to the absolute weight of B, so is line E to line F. Hence, by equidistance of ratios, the absolute weight of A is to the absolute weight of B as line D is to line F; which was to be proved. We may now pass on to show how:

If a solid cylinder or prism of less specific weight than water is [75] placed, as above, in a vessel of whatever size, and water is then poured in, the solid will remain unraised until the water arrives at that place in its height to which the entire height of the prism has the same ratio which the specific weight of water has to the specific weight of that solid; and, any more water being poured in, the solid will be raised.

Let there be the vessel MLGN, of any size, in which is placed the solid prism DFGE of less specific weight than water; and whatever be the ratio of the specific weight of water to that of this prism, let height DF have [that ratio] to height FB. I say that if water is poured in to height FB, the solid will not be raised, but it will be brought into equilibrium so that any little additional water added will raise it.

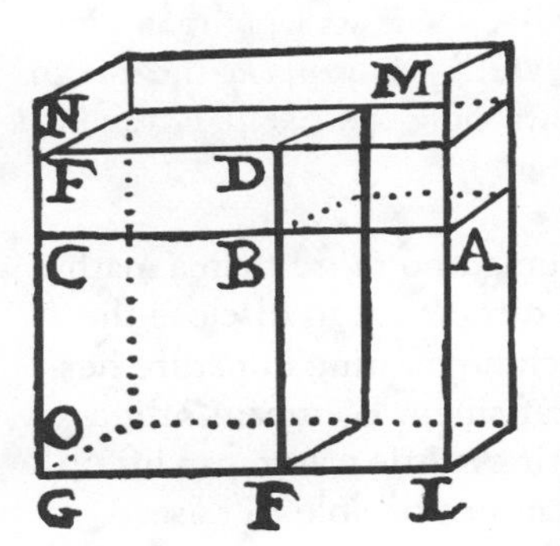

Let the water be poured in to the level ABC; then, since the specific weight of solid DG is to the specific weight of water as height BF is to height FD (which is as volume BG to volume GD), and the ratio of volume BG to volume GD and that of volume EF compounded together give the ratio of volume BG to volume AF, therefore the volume BG has to the volume AF the ratio compounded from the ratios of the specific weight of solid GD to that of water, and of the volume GD to the volume AF. But the same ratios (of specific weight of GD to that of water, and of volume GD to volume AF) also compound (by the foregoing lemma) to the ratio of the absolute weight of solid DG to the absolute weight of the volume of water AF; whence as volume BG is to volume AF, so is the absolute weight of solid DG to that of a volume of water AF. But as volume BG is to volume AF, so is the base of prism DE to the surface of the water AB, and the descent of water AB to the ascent of solid DG. Therefore the descent of water is to the ascent of the prism in the same ratio as that of the absolute weight of the prism to the absolute weight of the water, so that the moment resulting from the

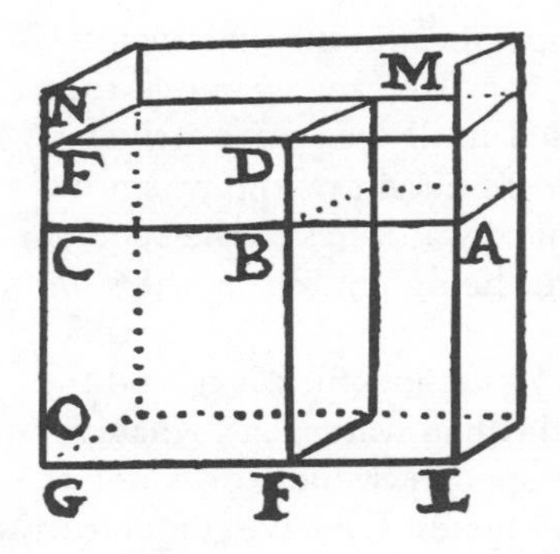

absolute weight of water AF and its speed in descending, with which moment it exerts force to drive out and raise up prism DG, is equal to the moment resulting from the absolute weight of prism DG and the speed of motion with which it, raised, would ascend, and since these two moments are equal [in this position], equilibrium will be reached between water and solid. It is then manifest that upon the addition of a little water to that at EF, its weight and moment will be increased, whereupon the prism

[76] DG will be overpowered and lifted until only the part BF remains submerged; which was to be demonstrated.

Salviati Now you see, Simplicio, how what seemed no more than a mathematical investigation of minutiae has turned out to disclose the explanation of a kind of engine or machine existing in nature herself,† not one devised by ingenious craftsmen. By means of it a small boy, merely by the effort of pouring a little water, can lift up a mighty beam that two strong men might not be able to raise. Knowledge of this fact may in turn enable ingenious men to devise more powerful engines than any we now have. It is true that such investigations, which do not compare with the achievements of philosophy, were considered by Plutarch (and other wise men) to be mean and unworthy, but you see why we need not believe that Archimedes was ashamed of them, or that our friend should be, or even that philosophical principles must originally have inspired their interest in such studies. And, since such trivial conclusions do not compete with philosophy, it is wrong for philosophers to oppose them and even declare them to be false and damaging to Aristotle's philosophy of nature.

Simplicio Nevertheless, such studies distract the attention and corrupt the understanding of men who might otherwise have been sound philosophers, abasing their minds from contemplation of higher things. That was already evident in our friend soon after he had got hold of the telescope and imagined he saw mountains on the polished surface of the celestial moon. Now he will make a mere machine of water, instead of devoting his acute mind to the place of that element in the scheme of the universe. And who knows where such speculations may lead next, if they go on unhindered?

Salviati I cannot answer for philosophers, but where our author was next led is shown by this, in which the famous conclusion of Archimedes is seen to follow from new principles:

From what has been shown, it is manifest that solids of less specific weight than water submerge only to a point such that water in the same volume as the submerged part will weigh as much as does the entire solid. For assuming that the specific weight of water has to the specific weight of prism DG the same ratio as that of height DF to height FB (which is as the solid DG to the solid GB), we shall easily prove that the same volume of water as that of solid BG weighs absolutely as much as the whole solid DG.

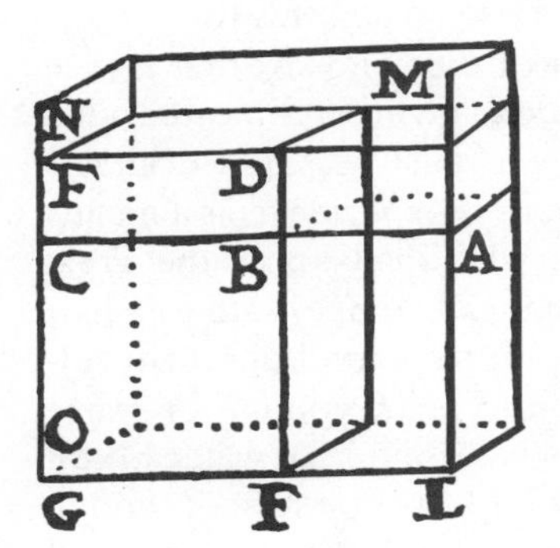

By the preceding lemma, the absolute weight of a volume of water equal to the volume BG has to the absolute weight of prism DG that ratio which is compounded from the ratios of volume BG to volume GD and of the specific weight of water to the specific weight of the prism. But the specific weight of water is assumed to be to the specific weight of the prism as volume DG is to volume GB; therefore the absolute weight of a volume of water equal to the volume BG has, to the absolute weight of solid GD, that ratio which is compounded from the ratios of volume BG to volume GD and of volume GD to volume GB—which is to say, the ratio of equality. So the absolute weight of prism BG is equal to the absolute weight of the whole solid DG.

It follows next that if a solid of less specific weight than water is placed in a vessel of any size, and water is placed around it to that height at which the volume of water equivalent to the submerged part of the solid weighs absolutely as much as the whole solid, that solid will be sustained by that water whether the water surrounding it is of great or small quantity.

For if the cylinder or prism M, less heavy than water in, say, the ratio of three to four, be placed in the large vessel ABCD and then water is raised around it to three-quarters of its height (that is, to level AD), it will be sustained and equilibrated exactly; and the same would happen if the vessel were very small, as ENSF, so that

[77] between it and the solid there remained a very narrow space, capable of holding hardly as much water as one-hundredth of the volume M, by which M would similarly be raised when this [the water] first arrived at three-fourths the height of the solid. At first glance this might seem to many a great paradox, giving people the idea that the demonstration of any such effect must be sophistical and fallacious. For those who so believe, there is experience as a means that may convince them. But anyone who shall know the great import of speed of motion, and how that can compensate to a hair any deficit or lack of heaviness, will cease to marvel upon considering how, on the raising of solid M, the great volume of water ABCD sinks very little, while the very small volume of water ENSF drops greatly and in an instant as the solid M is first raised, though through a very small distance. Hence the moment compounded from the small absolute weight of water ENSF and its very great speed in descending will equal the power and the moment resulting from compounding the great weight of water ABCD with its very great slowness in going down, inasmuch as in raising the solid M, the lowering of the little water ES occurs as much more swiftly than the great quantity of water AC moves, as the one

[quantity] is greater than the other. This is demonstrated as follows:

In the lifting of the solid M, its rise has to the fall of the surrounding water ENSF the same ratio that the surface of that water has to the base area of that solid M, which base has the same ratio to the water surface AD as the descent of water AC has to the rise of solid M. Hence, by perturbed proportionality, in the raising of solid M, the descent of water ABCD has, to the descent of water ENSF, the same ratio that the water surface EF has to the water surface AD; that is, [the ratio] that the whole volume of water ENSF has to the whole volume ABCD, they being equally high. It is evident, therefore, that in the driving and lifting of solid M, the speed of water ENSF exceeds that of water ABCD by as much as the former water is exceeded in volume by the latter, whence their moments in this operation are equalized.

Simplicio Excuse me, but what does he mean by "perturbed proportionality?"

Salviati That is a term introduced by Euclid in the fifth book of his *Elements,* dealing with ratios and proportionalities of continuous magnitudes, which would include speeds or volumes of water. When the first two terms of three proportional magnitudes are in the same ratio as the second two terms of another three, and the second two terms of the first have also the ratio of the first two terms of the second, then the ratio of the first and third terms is the same in both, and this is called *perturbed proportionality.*

Now ensues a remarkable passage inserted in the second edition to illustrate this noteworthy equalization of moments in a striking instance of a small weight of fluid balancing one much greater:

For a more ample confirmation and a clearer explanation of the same thing, consider the next diagram, which, if I am not mistaken, will serve to remove from error some practitioners of mechanics who at times attempt impossible undertakings on a false founda- [78]

tion. Here the large vessel EIDF is continued into the small tube ICAB, and we suppose water to be infused in these to the level L–GH. In this state it will rest, to the wonder of some people who do not immediately understand how it can be that the heavy load of the large volume of water GD, pressing down, does not raise and drive up the little quantity contained in the tube CL, which fights it and impedes such descent. But the marvel will cease if we begin to imagine the water GD to have dropped only as far as QO, and then consider what the water CL shall have done. To make room for that other water, which dropped from the level GH to the level QO, it will have to have been raised in the same time from level L all the way to AB, and the rise LB will be as much greater than the drop GQ as the area [*larghezza*] of the vessel GD is greater than that of the tube LC, which in short is as much as the water GD exceeds that of LC. But since the moment of speed of motion in one moveable compensates that of heaviness in another, what will be the wonder that the very swift ascent of the little water CL resists the slow descent of that great amount DG?

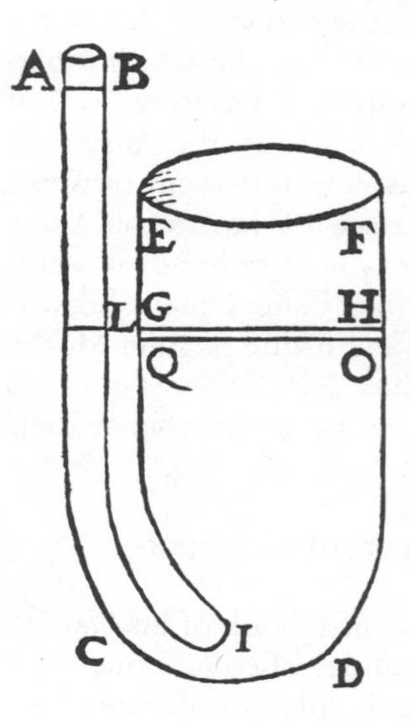

Simplicio I like this much better than those "compound ratios" and "perturbed proportionalities" of the previous demonstrations. But I fail to see why time enters into the matter. I meant to raise this question earlier.

Salviati Time comes in because it is inseparable from speed. Here the equality of times in which water would have to descend but a fraction of an inch in one vessel, while it rises many inches in the other, creates precisely that difference of speeds which just now made the whole situation clear to you. Equal time was also the basis of our friend's appeal to speed in a case that earlier caused you to object that since in equilibrium there is no motion, neither is there any speed. Here we are asked to imagine the consequences of a very small motion, which perhaps you will allow as a potential motion, or a virtual motion, and this helps us to understand why in this case motion does *not* become actual—because of compensating speeds and weights in what would have to be moved in order *not* to have equilibrium.

Simplicio I am perplexed, and must think it over. To speak of a "potential motion" contradicts Aristotle's very definition of motion, which is actualization of potential. I should resist this, yet at the same time I perceive the force of our friend's argument *ex suppositione.* This is, I suppose, good mechanics but bad physics, as might be expected in such mixed or bastard sciences as introduce mathematics into material considerations where it has no proper place. Why not just say, as good physicists do, that water seeks its own level, and account thus for the supposed marvel quite simply by an unquestioned principle?

Sagredo That seems to me a dangerous principle for the actual world, though splendid for a world on paper, being so simple and so clear.† In the real world, things float partly in water and partly out of it, which would not be the case if water did universally seek its own level. Ships would be quite differently designed if they had to slide along level water, like stones on ice.

Salviati Because principles favored by philosophers have seldom accorded with mechanics in practice, our friend was induced to seek other principles in mathematics, not without some success. He believes that better understanding may result, though for some people not simpler understanding; for as we have just seen this will require study of the *Elements* of Euclid. In any event, the first edition continued as follows, referring not to the connected vessels above, but to the next previous demonstration:

In this operation [of a solid raised by surrounding water] there therefore occurs exactly what happens in the steelyard. There a two-pound counterweight balances 200 pounds whenever, during the same time, the former must move through 100 times as much space as the latter—which is when one arm of the balance is one hundred times as long as the other. Accordingly this finishes off the false opinion of those who think that a ship is sustained better and more easily in a vast bulk of water than in a lesser amount (as was believed by Aristotle in his *Problemata* 23:3). On the contrary, it is truly possible that a boat should float as well in ten barrels of water as in the ocean itself.

Simplicio It pains me that our friend is forever contradicting Aristotle, who in that passage was proposing a problem rather than asserting a fact.

Salviati He intended rather to excuse others, as even Aristotle believed this; for he did not cite Aristotle in the first edition.

Sagredo But the problem was fictitious if the thing to be explained was not a fact, as I can assure you is the case. Had Aristotle been as guarded in asserting the supposed fact as he was in proposing a solution of the problem, he would not have exposed himself to such contradiction. I, who have sailed a great deal on the Mediterranean, have never seen the purported effect, and I do not believe it exists.

Salviati Our friend once remarked to me that, since Aristotle was of a probing mind, he probably sometimes attempted to solve problems that originated not in his own observations but in the tales of others. In this case Aristotle had probably heard from sailors that in harbor a ship may sink deeper into water than it does in the open sea, and supposed the effect to be universal.† But it occurs only in harbors near the mouths of rivers, where a ship sails in from the salty sea that buoyed it up more than does fresh water. Next:

But, getting back to our subject, I say that from what has been demonstrated thus far we can understand how any of the solids we have been discussing, if specifically heavier than water, could never be sustained by any quantity of water. For we have seen how the moment of a solid that is specifically as heavy as water, opposed by the moment of any volume of water, is capable of holding it [anywhere] up to total submersion without its being lifted, from [79] which it is manifest that much less could a body be raised up by water when it is heavier specifically than water. Then, being plunged in water up to total submersion, it will rest on the bottom, with such heaviness and opposition to being lifted as is the excess of its absolute weight over the absolute weight of a volume equal to it and made of water, or of material equally specifically heavy with water. And, although there should then be added some great quantity of water over the level covering the height of the solid, there will be no increase in any pressing or weighing down of the surrounding parts on the said solid, by which greater pressure it would have to be driven—because opposition [holding it still] is made only by those parts of the water that would also have to be moved on any motion of the solid, namely, those [parts] alone that are included between the two parallel surfaces that comprehend the height of the solid immersed in the water.

Sagredo I had never thought of it that way, but that certainly does clear up the question, showing why no matter how much water is added above the top of the submerged solid, there is no gain in pressure fitted to raise it.

Now, gentlemen, it is getting a bit late, and clearly we shall have to spend at least another afternoon or two on this book. Hence I suggest that we stop our reading at this point. Let us now go into Venice, admire its beauties, and dine there.

Salviati If Simplicio agrees, nothing will suit me better, especially as we have now finished with the first exposition and are ready to turn tomorrow to some philosophical objections and the replies to them. Simplicio, will you take up the reading next? For the text turns to topics more congenial to your studies than perhaps the foregoing have been.

Simplicio I shall be most happy, and if you will lend me the book I may prepare myself tonight.

Sagredo Then let us summon a gondola and put these matters of floating on water to practical test during the cool evening hours that lie ahead of us.

THE SECOND DAY

Salviati While we are waiting for the worthy professor, Sagredo, I might tell you that the part of our friend's book we read yesterday arose not from the dispute that took place shortly after his return to Florence, but from another three years earlier. At that time the celebrations attendant on the wedding of Prince Cosimo, now grand duke, required that an esplanade be built on the Arno, the construction of which was debated among various engineers.† Our friend being present on a visit, he expressed opinions adverse to one of the plans, which was based on the idea that concave and flat shapes tied together could be utilized to bear up greater weights than any other arrangement of floating supports. The more recent dispute also came to center on the role of shape in floating, though in a very different sense. This recalled to him the previous argument, in the course of which he had investigated the reason why large solids may float in small amounts of water. Finding that no one had previously explained this, he took occasion to set it forth in this book as a preliminary to the explanation of another and quite different phenomenon of floating. In the earlier dispute his adversaries were led by a military engineer, Don Giovanni de' Medici, who also entered the more recent debates, though those primarily involved philosophers rather than practitioners. Having been able to persuade engineers of the correctness of that basic cause of floating or sinking set forth in antiquity by Archimedes, our friend was the more astonished at the great difficulty of convincing philosophers, who are supposed to be the most willing of all men to yield to reason. In fact he found that a stubborn insistence on imagined causes of floating prevented their perceiving the true one, and to make any headway he was obliged to have recourse to many ocular demonstrations against their positions and favoring his own. I mention this because we are about to hear some arguments of quite a different character from those of yesterday, directed against a certain philosopher who was one of our friend's professors at Pisa when he was first imbibing natural philosophy as taught by Aristotle. Lest these seem out of place in a book of demonstrative science, I thought you might like to know the reason of their presence, as also of similar digressions later on.

Sagredo Thank you for telling me. When this book first appeared, many here thought it full of paradoxes and irrelevancies, though before long most changed their views and came to agree with our friend's positions and explanations.† Simplicio, I fear, held stoutly to Aristotle throughout, and without studying the demonstrations he inveighed against the use of mathematics in the quest for physical principles. But here he comes now, and we may resume. Greetings, Simplicio.

Simplicio Good morning, gentlemen. Our tour and supper last night left me time to do little more than scan the next few pages, which I see touch on issues more important, and more familiar to me, than those that occupied us yesterday. Shall I begin reading?

Salviati Please do, and thank you for relieving me of that obligation for a time.

Simplicio Starting from where we left off yesterday, the author reviews what has been done before turning to more profound arguments, thus:

It appears to me that up to this point there has been sufficiently described and opened a road to the contemplation of the true, intrinsic, and proper cause of the diverse motions and of rest of different solid bodies in various mediums, and especially in water, showing how in effect the whole thing depends on the interworking excess of heaviness of the moveables and of the mediums. And, what is most important, the objection has been removed that might for many people raise great doubts and difficulties concerning the truth of my conclusions—that the excess of heaviness of water over that of the solid placed in it is the cause of its floating and of its rising from the bottom to the surface, and that a quantity of water that weighs less than ten pounds may raise a solid weighing more than a hundred. For we have demonstrated that it suffices that such differences be found between the specific heaviness of mediums and moveables, the particular and absolute weights being whatever you please, in such a way that provided only that a solid be specifically less heavy than water, though in absolute weight it were a thousand pounds, it may be raised by ten pounds of water or less; while on the other hand a different solid, provided that it be specifically heavier than water, cannot be raised from the bottom and sustained by the whole sea. It suffices me, for what pertains to the present business, to have revealed and demonstrated this by the examples explained already, without further extending such material into a long treatise, as might be done—and indeed, had it not [80] been necessary to resolve the question just mentioned, I should have stopped merely at what Archimedes proved in the first book of his *On floating bodies,* where the same conclusions are reached in general and it is established that bodies less heavy than water float, those heavier go to the bottom, and those equally heavy stay indifferently at any place, provided they be completely under water.

But that doctrine of Archimedes having been read, transcribed, and examined by Signor Francesco Buonamico† in the fifth book of his book *On Motion,* at chapter 29, and there confuted by him, it may be rendered dubious and suspect of falsity by the authority of so celebrated and famous a philosopher. Hence I judge it necessary to defend this doctrine, if I shall be able to do so, and to exculpate Archimedes from the faults charged against him.

Buonamico abandons the doctrine of Archimedes in the first place as not in accordance with the opinion of Aristotle, adding that it seems to him astonishing that water must be [deemed] heavier than earth when on the contrary it is seen that heaviness in water increases through its mixing with earth. Next he remarks that he remains unsatisfied with the reasons of Archimedes because with that doctrine he cannot give the cause why a boat and a vase, which otherwise stay afloat on water, go to the bottom when filled with water; since the weight of the water contained therein is equal to that of the other water and ought to come to rest at the surface, yet it is seen to go to the bottom. Moreover, he adds, Aristotle clearly refuted those ancients who said that light bodies are moved upward as driven by the impulsion of the heavier surroundings, for if that were so it should follow that all physical bodies were by nature heavy, and none light, because the same would happen to air and to fire if placed at the bottom of water. And, although Aristotle grants an impulsion in the elements by which the earth is reduced to spherical shape, in his [Buonamico's] opinion this impulsion is not such as could remove heavy bodies from their natural place; rather, it sends them toward the center, to which (as he somewhat obscurely goes on to say) water also principally moves unless it encounters something that resists it and cannot be driven from its [81] place by the weight of the water. Thus, if not directly, at least however it can, water goes on to the center, and it is only accidentally that light things go upward by this impulsion, they having this [upward] tendency by their nature, as also to remain floating. Finally he concludes by agreeing with the conclusions of Archimedes, but not as to the causes, which Buonamico wishes to refer to the easy or difficult division of the medium and to the dominions of the elements, so that when the moveable overcomes the power of the medium (as, for example, lead overcomes the continuity of water) it will move through that, and otherwise not.

Sagredo If I may interrupt for a moment, it seems to me that all this is much ado about nothing, since at the end we learn that this opponent of Archimedes agrees with all his conclusions and would quarrel only over the means by which he arrived at them. This is as if two generals, both agreeing that an assault should be made at once on a certain place, but for different reasons, should stop to argue which reasons were right and thereby lose the battle by giving the enemy time to fortify that place.

Simplicio No indeed, Sagredo; in science all that really counts is the understanding of causes and reasons. The consequent facts, which are all we are given in conclusions, are of little concern to the natural philosopher, those being in the province of the mere practitioner. Thus far it appears to me that Professor Buonamico speaks sagely and is entirely in the right, he having destroyed the reasons of Archimedes, though accepting his conclusions, and in their place having put this clear physical principle of resistance to division, which the Syracusan seems entirely to have overlooked. It is a very grave defect in any science to omit true causes and offer defective ones, by our author's own admission yesterday. But I said "thus far," for we must be fair and give our friend his chance to defend Archimedes if he can—though that will be hard, against so powerful an opposition.

Sagredo My opinion is still that his time would be better spent in going on to new and useful conclusions, since those already established went unchallenged, leaving to philosophers all disputes over causes, as astronomers have done ever since the time of Geminus. But if he wishes to put aside science and play the philosopher, I am more than ready to listen, especially as I think few philosophers living would attempt to support the conclusions of Archimedes, or even to understand them, leaving that to mechanicians and mathematicians.

Simplicio You are right there, for even by the most convinced Platonists of my acquaintance I have never heard a good word said for this book of Archimedes, or any word at all—they holding that we can never learn truth about the real world by frittering away time on this poor world of illusions and appearances. Aristotelians, on the other hand, might come to the support of Archimedes were it not for the fact that he, by putting mathematics in place of Aristotle's principles of nature, failed to provide any true cause for his conclusions.† But let us read on:

Such is what I have been able to gather to have been produced against Archimedes by Signor Buonamico, who did not take the trouble to challenge the principles and suppositions of Archimedes, since those would necessarily be false if the doctrine dependent on them is false; so he was content to adduce some contradictions and inconsistencies with Aristotle's doctrine. To these objections I reply by saying first that the simple fact of discordance between the doctrine of Archimedes and that of Aristotle ought not to move anyone to hold the former suspect, there being no cause whatever for which the authority of one need be placed above the authority of the other.† Where we have decrees of nature exposed indifferently to the intellectual eye of everyone, the authority of either man loses any persuasive force, the absolute power remaining with reason.

Therefore I pass to what was adduced second—that it absurdly follows from the doctrine of Archimedes that water must be heavier than earth. Now, truly I do not find that Archimedes said any such thing, or that it can be deduced from his conclusions; and if that were made manifest to me I believe I should absolutely abandon his doctrine as most false. Perhaps Signor Buonamico supports this deduction upon what he adds about the vase that floats when devoid of water but goes to the bottom when filled;

and, supposing the vase to be an earthen one, reasons against Archimedes thus: "You say that solids which float are less heavy than water; this earthen vase floats; therefore this vase is less heavy than water, and so earth is less heavy than water." If that is the train of reasoning, I easily answer by granting that such a *vase* is less heavy than water, but denying the second consequence, that *earth* must then be less heavy than water. The floating vase occupies in water a place equal in volume not only to the earth of which it is made, but [82] equal to that earth together with the air contained in its hollow; and if such a volume composed of earth-and-air shall be less heavy than as much water, it will float on top, and this will conform to the doctrine of Archimedes. But if we then remove the air and fill the vase with water, so that the solid placed in water shall be nothing else but earth, it does not occupy any place other than that filled by that earth, and then it will go to the bottom because earth is heavier than water; and this agrees well with the opinion of Archimedes.

Here is the same effect, explained by another similar experience. In trying to push to the bottom a glass tumbler when it is full of air, great resistance is felt, because it is not just the glass that is pushed under water, but together with the glass a great bulk of air, and such that if you took as much water as is the volume of the glass and of the air contained in it, it would have much greater weight than do the tumbler and its air, which therefore will not submerge except with great force applied. But if you put in water just glass, as will be the case when the tumbler is full of water, then the glass will descend to the bottom as a thing greater in heaviness than water.

Returning now to the original proposition, I say that earth is heavier than water, and that therefore an earthen solid goes to the bottom; but a composite of earth-and-air can indeed be made that will be less heavy than the same volume of water, and it will remain afloat; and both experiences agree very well with the doctrine of Archimedes. But since it appears to me that there is no difficulty in this, I do not want positively to affirm that Signor Buonamico wished by such reasoning to oppose [as if] against Archimedes the absurdity of inferring from his doctrine that earth was less heavy than water—though truly I cannot imagine what other event may have induced him to do this.

Simplicio The next brief paragraph was inserted in the second edition here:

Perhaps it was a question (or fable, in my opinion) read by Signor Buonamico in another author that led him to attribute a singular property [of floating earth] to some particular water; here this comes to be used, doubly in error, to refute Archimedes, who said no such thing. Nor did he who told the story [of a supposed floating island] assert that the water [on which earth floated] was our common element.

The third difficulty in the doctrine of Archimedes was that it [allegedly] could not give a reason why it happens that a boat—and a vase, even of wood—that will otherwise float, goes to the bottom when full of water. Signor Buonamico believed that a wooden vase, [83] of wood that naturally floats, goes to the bottom promptly on its being filled with water, and speaks at length of this in the next chapter (which is the thirtieth of the fifth book). But I, speaking always without belittling his great learning, shall dare in defense of Archimedes to deny any such experience, being certain that wood which does not by its nature sink to the bottom in water will not do so when carved in the form of any vase whatever and then filled with water. Anyone who wants quickly to see this [by] experience with some flexible material that is easily shaped to any figure can take some wax and, first making of it a ball or some other solid shape, add to it as much lead as just barely makes it sink, so that one grain less would not suffice to make it submerge. For then, giving this the form of a vase and filling it with water, one will find that without the same lead it will not go to the bottom, and that with the lead it will very slowly descend; and in a word one may ascertain that the contained water has no effect whatever.

I do not deny that one can make, using wood that naturally floats, boats that will sink when filled with water; but that happens through weight gained not from the water, but from the nails and other ironwork, so that one no longer has a body less heavy than water but a composition of wood-and-iron heavier than an equal volume of water. So let Signor Buonamico desist from trying to render reasons for an effect that does not exist. Indeed, if this "sinking of a wooden vase full of water" could throw doubt on the doctrine of Archimedes, according to which it should not sink, and on the other hand this agrees with the Peripatetic doctrine, which could commodiously assign a reason that the vase should submerge when full of water, we might turn the argument around and say with certainty that the Archimedean doctrine was true because it closely fitted true experiences, and the other doubtful, since its deductions accommodate false conclusions.

Then, as to the other point mentioned in this same objection, where Signor Buonamico seems to say the same not only of [ordinary] wood shaped in the form of a vase but also of dense wood, which I believe he means to be soaked and impregnated with water, which goes finally to the bottom [I say] that happens with some porous woods that, when they have the porosities full of air or other material less heavy than water, are bulks specifically less [84] heavy than water, just as is that tumbler of glass when it is filled with air; but when that lighter material departs and water succeeds it in the porosities and cavernous places, it may very well be that then the composite remains heavier than water—just as, when the air leaves the glass tumbler and water enters, there results a composition of glass-and-water heavier than the same volume of water. But the excess of weight is in the material of glass, and not in the water, which is not heavier than itself; and thus the remainder of the wood, if that remainder shall be specifically heavier than water when the air leaves its concavities, and its porosities are filled with water, will make a composite of wood-and-water that is heavier than water—in virtue not of the water received in the porosities, but of that [heavier] wooden material which remained when the air departed; and hence it will go to the bottom in conformity with the doctrine of Archimedes—just as before, by the same doctrine, it floated.

Finally, as to that which was opposed in the fourth place—that is, that Aristotle had already refuted the ancients who, denying positive and absolute lightness, and holding that truly all bodies are heavy, said that what is moved upward is driven by the ambient, wherefore the doctrine of Archimedes, as an adherent of that position, was also overthrown and refuted [by Aristotle]: it appears to me that Signor Buonamico imposes on Archimedes, deducing from what he said more than he set forth or than can be deduced from his propositions, inasmuch as Archimedes neither denied nor admitted positive lightness, nor dealt with it at all, wherefore much less should anyone infer that he denied that it might be the cause and principle of upward motion of fire or of other light bodies. He merely, having proved that solid bodies heavier than water descend in it according to the excess of their heaviness over its heaviness, likewise demonstrated how solids less heavy ascend in the same water according to its excess of heaviness over theirs. Hence the most that can be deduced from the demonstrations of Archimedes is that just as excess of weight of the moveable over weight of water is the cause of its descent therein, so excess of the weight of water over that of the moveable is enough to prevent its descent and even to make it float, without [making] any inquiry whether upwaard [85] motion is or is not a further cause contrary to heaviness. Archimedes reasoned no less properly than one who might say: "If the southward wind strikes the boat with greater impetus than the force of the flow of the river carrying it northward, its movement will be southerly; but if the impetus of the water shall prevail over that of the wind, the movement will be northerly." The reasoning is good, and it would be unworthily attacked by an opponent saying: "You improperly adduce as the cause of the ship's northward motion that impetus of the flow of water in excess of the force of the southward wind—incorrectly, I say, because the force of northward wind, contrary to southward wind, [also] has power to drive the ship northward." Such an objection would be redundant, because he who adduced the flow of water as the cause of motion did not deny that a wind contrary to the southward wind could have the same effect; he only affirmed that when the impetus of the water prevails over the force of the southward wind, the ship will move northward, and that is truly said. Just so, when Archimedes says that, by the heaviness of water prevailing over that by which the moveable goes downward, such a moveable is raised from the bottom to the surface, he induces a very true cause of that event; nor does he affirm or deny that there is, or is not, a force contrary to heaviness (which some call "levity") that is also capable of moving some bodies upward.

Therefore the weapons of Signor Buonamico are directed [only] against Plato and other ancients who, completely denying "levity" and holding all bodies to be heavy, said that movement upward is made not by some intrinsic principle of the moveable but only by the driving of the medium, while Archimedes with his doctrine remains untouched, he having given no cause for impugnation.

But lest this excuse adopted in defense of Archimedes appear to some too weak to free him from the objections and arguments of Aristotle against Plato and the other ancients, and they might again rise up against Archimedes and those who adduce the driving of the water as the cause that solids less heavy than water [though not absolutely light] return to float, I would not shrink from sustaining as very true the opinion of Plato and those others who absolutely deny "levity" and affirm that in elemental bodies there is no other intrinsic principle of movement except toward the center of the earth, nor any other cause of upward motion (meaning for that which appears to have natural motion) than the driving of the fluid [86] medium that exceeds the heaviness of the moveable. I believe that Aristotle's reasons to the contrary can be fully satisfied, and I would undertake to do this if it were really necessary to the present subject, or would not make too long a digression in this short treatise. So I shall say only that if in any of our elemental bodies there were an intrinsic principle and natural inclination to flee from the earth's center and move toward the orb of the moon, any such bodies would doubtless ascend [even] more swiftly through mediums that less greatly oppose the speed of the moveable through them. Now, such mediums are the most tenuous and thinnest, as for example air in contrast with water; this we prove every day, since we much more easily move our hand, or a board, back and forth speedily in air than in water. Yet no body whatever is found that does not ascend much faster in water than in air, while of those bodies that we continually see move upward speedily in water, there is none that does not lose all its motion when it reaches the boundary of air. This [rule] extends to air itself, which, rising swiftly through water, loses all impetus on coming to its own region and slowly mixes with the rest. And inasmuch as experience shows us that bodies successively less [specifically] heavy move upward in water more swiftly, one cannot doubt that fiery exhalations ascend through water even faster than air does, while air is seen by experience to ascend more swiftly through water than fiery exhalations do through air. Hence one necessarily concludes that those same exhalations ascend much more swiftly through water than through air, and that consequently they are moved by being driven by the surrounding medium, and not by some intrinsic principle in them of escaping from that center toward which the other heavy bodies tend.

Simplicio I have read thus far without stopping because it was only fair to complete this argument against Aristotle before examining it, and here our friend changes to a new subject. Now, although he remarked that the question whether there is anything having "levity" or absolute lightness (which Aristotle says exists, while Plato and others denied this) is not germane to his discussion, he did take sides against Aristotle on this matter, which is of such vital importance to the whole of physics that I must reply.

If we were to allow that only heaviness, and not lightness, exists among elemental things, the whole doctrine of natural places, and along with it all explanation of natural motions, would be called into question, leaving physics with no foundation whatever. I am willing to grant that these trifling problems of floating and sinking in water can be solved without my going into those more profound doctrines, and I shall not enter into them now, but I must say that they are not to be settled by arguing from relative speeds of motion in some observed things to cases impossible to observe, and probably incapable of existing. Such would be the rising of fiery matter through water, where fiery matter is never found. Sparks fly upward, but sparks do not ascend in water, and it is great folly to reason that fire is not absolutely light because fire would not rise more slowly through water than through air. Philosophers arrived at the principles of levity and gravity by induction from a myriad of phenomena, and those principles cannot be shaken by arguing from mere speeds, especially imaginary speeds never observed by anyone.

Salviati I am inclined to agree, Simplicio, that our friend should not have got himself involved in matters not germane to his purpose, as this was by his own admission. On the other hand, I might tell you that our friend has shown me a treatise of his on motion,† composed at Pisa before he came to Padua, in which he argued very cogently that nothing would be lost if the element of fire were considered not absolutely light, but only the lightest of the elements, showing that nevertheless the whole doctrine of natural places and natural motions could be preserved. For that, it is only necessary that the concept of relative weights and relative places be more adequately developed than in the past, something our friend proposes to do in some later book, he being now much dissatisfied with that early treatise of his for other reasons that we need not bring up now. So be assured that if he sometimes transgresses from science into the domain of philosophy proper, that is not done by him as thoughtlessly of consequences as you now have reason to believe.

Simplicio Very well, we shall say no more about this now; but please remember that questions as important as Aristotle's ascription of intrinsic properties to elemental matter are not even touched by the foregoing flippant argument. Now, to resume our reading:

To that which Signor Buonamico adduces as his ultimate conclusion, he wishing to reduce descent or no descent to ease or difficulty of dividing the medium and to the domination of the [various] elements, I respond as to the first that the former can in no way be the correct cause, inasmuch as in no fluid medium, such as air, water, and other moist things, is there any resistance whatever to division, but all are divided and penetrated by any the least force, as I shall demonstrate below. Hence no such "resistance to division" can act at all, since it does not even exist.

As to the other, I say that it is the same thing to consider in a moveable the dominant of [two or more] elements as it is to consider the excess or defect of heaviness in relation to the medium, [87] since in such action the elements operate only to the extent that they are heavy or light. So it is the same to say that fir does not sink because air predominates in it, as to say [it does not sink] because it is less heavy than water. Indeed, the immediate cause is its being less heavy than water, and the predominance of air is the cause of less heaviness, so that whoever offers as the cause the predominance of this element adduces the cause of the cause, not the proximate and immediate cause. Now, who does not know that the true cause is the immediate cause, and not the mediate?

Simplicio Really, this can never be allowed, though I do not mean to defend the mediate. But the true cause, far from being the immediate cause, is the final cause; that is, the ultimate reason for which things must be as they are and not otherwise, because it is best that they be so. What our friend here calls the immediate cause is no other than what philosophers call the efficient cause, which is so far from being the true cause that it is capable of being regarded as hardly more than an accident, an event that (so to speak) happens to occur in fact and permits the essential cause to exhibit its working when something had been lacking up to that time or had prevented action. When a heavy body is held above the earth and then released, we may say that release is the immediate cause of its falling to earth; but the true cause of its falling is that as a heavy body it seeks its natural place; or we might even say that the true cause is its heaviness. Its release as such no more causes it to go down than up, or sideways, and is called a "cause" only for convenience, as when we say "this fell *because* its support collapsed." There can be no science of immediate or effective causes, since those are particulars and may occur in any number of ways.

Salviati Well, Simplicio, we have suddenly arrived at an impasse if you insist on this manner of talking. As you will see in due course, our friend takes as the cause of any effect that which is always present when the effect is seen, and in whose absence the effect does not take place.† Hence we may as well abandon our reading and our discussions at this point if you cannot accept his definition, worthy of Hippocrates—for you must know that our friend also studied medicine.

Simplicio On such a central notion as that of cause, people are not free to adopt arbitrary definitions at will. For science is understanding in terms of causes, and there can be no science if true causes are abandoned.

Salviati Certainly our friend would not advocate the abandonment of causes in his science, any more than in medicine. Not only does he use the word very often, but in many instances he implies it when the word itself is not present. It is a word of the greatest use in every act of reason, and is perhaps inseparable from reason, as suggested by the close resemblance of our words *cagione* and *ragione;* frequently it is immaterial whether we offer the cause of a thing or the reason for a thing. But in order not to have to cease and desist from our reading and discussion, which I for one find most interesting, let us compromise in this way. We can accept for our present purposes the medical and perhaps unphilosophical conception of "cause" that our friend has adopted, since he has given it a clear definition in terms of absence or presence together with an effect, and we can each of us preserve whatever mental reservation we please concerning its value, its completeness, and the limitations it places on the science he offers with or without proper "understanding in terms of causes." His definition will then stand as a kind of hypothesis, whose implications we can determine without being committed to its eventual acceptance. Is that agreed, at least for a time?

Simplicio I must admit that I too am enjoying our discussions, so for the moment I agree and shall try to understand him in this way, reserving the right to intervene when I think this may lead us astray. So, to resume the reading:

> Besides, he who alleges heaviness brings forth a cause well known to our senses, because we can very easily ascertain whether ebony, for example, or fir, is heavier or less heavy than water; but who will make manifest to us whether the element of earth, or that of air, has predominance in them? Certainly there is no better experience of this than to see whether they float or go to the bottom. So that whoever does not know that such a solid floats unless he [first] knows that air predominates in it, does not know that it floats until he sees it float. For he knows it floats when he knows air has predominance, but he does not know that air predominates except when he sees it float, and therefore he does not know that it floats except after having seen it float.

Sagredo I should like to remark here that, when I first read this, I thought it a sophistical bit of logic-chopping unworthy of our friend. But that bothered me, and now I have come to see this as a most profound observation about nature, philosophy, science, and the very foundations of science. It seems to me that he is saying that we possess true knowledge about effects in nature only if we have two different observations, of different kinds, and we possess only empty words when we have but one kind of observation and a logical consequence thereof. Thus in this case we have one way of finding the weight of fir in relation to that of water, by weighing equal volumes; and we have another way of finding out whether fir floats in water, by placing it there. But the adversary has only the latter, to which his logical deduction that air predominates over earthy stuff in the composition of fir adds naught, for he has but one kind of observation to which he gives two names. I may not have put the matter clearly, but what do you gentlemen think of it?

Salviati It is quite clear to me, and I can confirm your correctness in supposing that our friend was here not merely chopping logic, but putting forth something he has many times told me is always necessary in science, though often neglected in philosophy.

Simplicio I think you must be jesting, since science is a part of philosophy. Or is our friend so bemused by his mathematical speculations as to confuse them with science?

Salviati Not entirely, since he insists also on sensate experiences. But he has become interested in the pursuit of science in such a way that it may enable him to philosophize better,† as he puts it, and for that purpose this pursuit must be independent of philosophy, or at least of metaphysics, lest we end with but two names for one thing. He has found that sensate experiences are most useful when they permit measurements, to which necessary demonstrations may sometimes be found to apply, and in that way his principles of physics may be tested, refined, or replaced. But we are getting far afield, and in any event this discussion might better be postponed until we have heard all that is said in this book.

Simplicio Agreed. Well, next he says:

We do not deprecate those advances, therefore, though all too tenuous, that are made in our knowledge by reason after some contemplation; and we accept from Archimedes the understanding that any solid body will go to the bottom in water when it is specifically heavier than water, and that when it shall be less heavy it will necessarily float, while it would indifferently rest at any place in water if its heaviness were completely equal to that of water.

These things explained and established, I come to the consideration, in this matter of movements and rest, of the role played by differences of shape given to the moveable, and I again declare that:

Diversity of shape given to this or that solid cannot in any way be the cause of its going, or not going, absolutely, to the bottom, or of floating; so that a solid of, for example, spherical shape, that goes to the bottom, or floats in water, shaped in any other way—I say—will in the same water go down, or return from the bottom; nor will this motion of it be prevented or removed by [altering] the breadth, or by any other change of figure.

[88] Breadth of figure may indeed retard the *speed* of descent or of ascent, and the more as the shape is brought to greater breadth and thinness; but that shape can be altered so that it completely prevents that same material from moving any farther in that same water, I deem to be impossible. In this I have met with staunch contradictors who, producing some experiences, and in particular a thin chip of ebony and a ball of that same wood, showed how the ball sank to the bottom in water, while the chip, placed lightly on the water, did not submerge but stopped. They have deemed, and have been confirmed in their belief by the authority of Aristotle, that truly the cause of rest in this case is the breadth of shape, [making a body] unable by its little weight to fend and penetrate the resistance of the bodily character of water, which resistance is quickly overcome by the other, rounded, shape.

This is the principal point of the present question, in which I shall endeavor to make it clear that I am on the right side.

Therefore, commencing to investigate with examination by exact experiment how true it is that shape does not at all affect the sinking or not sinking of the same solids, and having already demonstrated how a greater or less heaviness of the solid with respect to the heaviness of the medium is the cause of its ascending or descending, [then] whenever we want to make a test of what effect diversity of shape has on the latter, it will be necessary to make the experiment with materials in which variety of heaviness does not exist. For were we to make use of materials that could vary in specific weight from one to another, when we encountered variation in the fact of descent or ascent we would always remain with ambiguous reasoning as to whether the difference derived truly from shape alone, or also from different heaviness. We may protect ourselves against this by taking a single material that is tractable and suitable to be easily reduced to any kind of shape. Moreover, it will be best to take a kind of material quite similar to water in heaviness, because such a material would be indifferent, so far as heaviness is concerned, to sinking and rising, whence it will easily be known what little difference may derive from change of shape.

Now wax is very suitable for this, since besides its receiving no sensible alteration from impregnation by water, it is tractable, and [89] the same piece is very easily brought to any shape; while being very little less heavy than water, it can be brought to very nearly equal heaviness therewith by imbedding in it a few lead filings.

Simplicio Here I begin to wonder how far we should trust the procedures described, when what is in question is the universal statement that no change of shape has any bearing on the sinking or rising of any material. It begins to appear that "any material" is going to turn out to mean "wax," and "any shape" to mean "the few shapes tested, out of an indefinitely large number of shapes."

Salviati That objection was raised at Florence by Professor di Grazia, who declared that in order to support a general principle by induction it would be necessary to examine not just a few shapes, but all shapes.† Our friend replied that in no way could that be the use of induction, since when the number of forms in unlimited, it could never be so applied, and when the number is limited and every single case is analyzed, induction is superfluous—or rather, the conclusion is not by induction. As you will see, certain shapes were selected—particularly thin flat chips with parallel surfaces, spheres, and cones or pyramids. Our friend contended that argument a fortiori was of special use in induction, since if an effect is seen to hold for a very unlikely shape, it is safe to assume the same thing for shapes more probable in relation to the particular effect being tested. Since we shall come upon some very apt examples of this, we may defer the whole matter until later.

But now, Simplicio, since we are entering on the matter of experiment in science and leaving the philosophical discussion for a time, perhaps you would prefer to have Sagredo take up the reading, freeing yourself to raise questions and objections as they occur to you. I take it that Sagredo will not object, and I may add that our friend has told me of Sagredo's special apritude for and interest in experimental activities.

Simplicio Thank you for suggesting this, which I hesitated to do, though you have exactly expressed my reasons. Do you mind, Sagredo?

Sagredo Indeed not; I have been feeling a bit left out, having had nothing to contribute to this part of the discussion because of my ignorance of philosophy and my agreement with our friend's statements. Now I may at least contribute by the reading of his text. He next says:

> Such material being prepared, make it for example into a ball the size of an orange or larger, and heavy enough to stay at the bottom, but so lightly that by the removal of a single grain of lead it will float, and then with that grain restored it will return to the bottom. Next, reduce the same wax to a thin broad leaf, and try the same experiment again. You will see that, placed on the bottom with that added grain of lead, it remains there, and with that grain removed it will rise clear to the surface; but with the grain of lead restored again, it will sink to the bottom. And this same effect will always take place in all sorts of shapes, regular as well as irregular, nor will any ever be found that will come to float without removal of the grain of lead, or will sink to the bottom without its addition.

Sagredo Here I shall interrupt to say two things. First, I call your attention to the relatively large size that was specified here by his saying that the body should be the size of an orange or larger. On that condition the above statements will be found to be quite correct, for—and this is my second statement—I have carried out the experiments described in this book, and from time to time I shall indicate what I found, for not all of them are equally easy, and in some I was not successful.†

Simplicio Well, if the rule holds only for large solids, our friend was seriously in error by not telling us so at the start. Even more reprehensible is the procedure I deduce from your second statement, for it is improper to base one's arguments on experiments that fail, treating them as if they had succeeded. I am astonished and grieved that our friend, whom I knew as at least honest, even if sometimes headstrong and rash in opposing Aristotle, would stoop to this.

Sagredo Easy now, Simplicio; I did not say either that his rule holds only for large bodies or that some of his experiments are fictitious. As to size, I called attention to a specific condition imposed by our friend on this particular introductory experiment. You will see presently that a special kind of floating may occur in objects below a certain size, which phenomenon differs with each material of different specific heaviness.

Salviati Our friend was more than aware of this, for the fact was forcibly called to his attention by his most vociferous opponent, Lodovico delle Colombe,† three days after the original discussions, at which this philosopher had not been present; or rather, I should say it was called to our friend's attention by Professor di Grazia then, as a result of a certain challenge made by Colombe. Now, in writing for the general reader, unfamiliar with the subject, our friend began by describing a certain specific experiment that precludes the newly introduced kind of floating—or better, as he put it, of staying atop water; for, as we shall see, what floats in such cases is not just what is seen to float and is vulgarly taken as the floating object. In that way he was able to present some initial statements borne out by test without becoming involved in the discussion of seeming exceptions, which he introduces later in the proper place. Here we have again that "for the most part."

Sagredo So much for the preliminary specification of a minimum size. As to failure on my part to obtain experimental agreement in some cases, that was not quite what Simplicio may now think. Anyone may fail to obtain the same results as another through careless-ness, or improper preparation of materials, or want of dexterity; but such failure does not mean that science cannot be erected on similar experiments. That one surgeon may save and another may kill his patient by performing the same operation, say removing a tumor, does not invalidate the surgical procedure. What I had in mind were some instances in which, experimenting with all the care I could muster, I obtained the expected result only sometimes, not always, and I became aware that at times tiny bubbles of air cluster on the wax and bring it to the surface, or keep it afloat, and removal of those bubbles (without adding a grain of lead) is followed by its sinking. Now, in some very delicate experiments I was unable to say whether the effect sought was truly observed or not, and by this I mean that the matter was so difficult that I could not be sure that our friend had not been deceived in the way I just mentioned. Knowing him as I do, I should hesitate to declare that he was deceived, in this way or in some other, for I think it improbable that I noticed anything he overlooked. Rather, I think it likely that he omitted from his book a good many precautions that would make the general reader simply impatient, and that would enable philosophers to say that nothing of value can be deduced from such finicky precision, much as Aristotle said that some people thought it undignified to seek mathematical precision in physics, likening this to haggling over prices in the market.

Simplicio I apologize for my hasty judgment, but I am nevertheless very glad that I made it, for your further explanation has not only removed my impression that our friend had resorted to deceit in order to persuade unwary readers, but has also given me reason to think further about the experimental procedures you describe. There may be something in them analogous to minute distinctions that philosophers have found to have great consequences in the end. I shall need time to ponder this before saying more, so please go on.

Sagredo Then this present digression has been profitable, though at first it seemed to threaten grave division between us. I know only too well, as probably Salviati does too, why you momentarily thought monstrous something that is a necessary part of our friend's style in introducing to his readers things entirely new to them. Time and again I have rebelled against his seeming paradoxes, only to learn that they were appearances born of my failure to attend to his exact words† and then to wait patiently for clarification of the whole knowledge that made him aware, from the outset, of some necessary qualifications. Now, to go on:

To sum up, no difference whatever is discerned [among different shapes] with regard to sinking or not sinking to the bottom. But it is not so concerning the swift and the slow [movement of the solid], since broader and more extended shapes do move more slowly, as well in dropping down as in rising, while narrower or more compact shapes move more swiftly. Now, I do not know what difference must accompany variety of shapes, if the most different among them act [on the motion] no more than [does] a very tiny grain of lead affixed or removed.

I think I hear some adversaries raise a question about the experiment I have produced, first demanding that I consider that mere shape may have no effect as simple shape, separated from any material, and yet must have [an effect] when conjoined with matter, though not with just any material, but only with particular material than can produce the desired effect. Thus we see it to be true by experience that a narrow acute angle is more fitted for cutting than an obtuse angle, provided, however, that both shall be conjoined with material capable of cutting, as for example with iron. For a knife with a thin acute edge cuts bread very well, and even cuts wood, as it would not do if the blade were thick and obtuse; yet whoever might wish to use wax instead of iron, making a knife out of wax, would truly not be able to recognize in that material what effect the sharp blade has, and what the obtuse one, since neither [90] one would cut, the softness of wax rendering it unsuitable to overcome the hardness of wood, or even of bread. Then, applying similar reasoning to our proposition, they will say that different shape does show a difference of effects concerning sinking or not going to the bottom, though not when conjoined with just any material, but only with those materials which by their weights are suited to overcome the resistance of the viscosity of water. Therefore anyone who should take as his material cork or some other light wood, unable by its lightness to overcome the resistance of the bodily character of water, and should form of such material solids of different shapes, would attempt in vain to see how shape affects sinking or not sinking, since all would remain afloat—not by some property of this or that shape, but by the feebleness of material lacking the necessary weight to overcome and vanquish the density or bodily character of water.

Therefore it is necessary, if we wish to see what differences of shape may effect, first to select a material able by its nature to penetrate the bodily character of water; and for that purpose they thought a material would be suitable that, whenever reduced to spherical shape, went to the bottom; and they chose ebony. Making then a little chip of this as thin as a vetch pod, they showed how this rests when placed on the surface of water, without sinking to the bottom, while on the other hand they showed that a ball made of the same wood and no smaller than a hazelnut does not stay afloat, but sinks. From that experiment they thought they could freely conclude that in the flat chip breadth of shape was the cause of its not sinking to the bottom, inasmuch as a ball of the same material, differing from the chip only in shape, went to the bottom of the same water. And truly the reasoning and the experiment have so much probability and verisimilitude that it would be no wonder if many, persuaded by a certain first appearance [of truth], should lend their assent to this; yet I believe I can show here no lack of fallacies.

Simplicio However this turns out, I must say that the philosophers did themselves credit by exhibiting the material cause and thus showing how deeply one must consider the properties of different materials in this matter of testing the effects of shape, and not just go generalizing from wax. This shows again that science can in no way be separated from its mother, philosophy, without grave risk to the child that would like to go exploring alone.†

Salviati What you say would be most true if in fact the mother acted as guide, leading the offspring to new glades and previously unseen flowers, but not if the mother protects the child by shutting him in a closet whenever he shows signs of boredom at observing her potted plants. I look forward to hearing what you may have to say further when you have seen how in this case the philosophers, having shown our friend something he had not explained, responded to his explanation and to the new things he revealed in turn—for which they themselves lacked any explanation.

Sagredo Then we must let his story unfold; for, as he remarked, truth reveals herself tranquilly and slowly:

Therefore beginning to examine bit by bit what has now been set forth, I say that shapes, as simple shapes, not only do not operate in physical things, but are never even found separate from bodily substances; nor have I ever proposed shapes denuded of sensible [91] matter. I also freely admit that in attempting to examine what differences in events there are that depend on variety of shapes, those must be applied to materials that do not impede the varied operations of varied shapes. And I admit and concede that I should have done badly had I wished to experiment on the role of sharpness of a blade with a wax knife, applying it to cut oak, because there is no sharpness whatever which, given to wax, will cut very hard wood. But it would not be inappropriate to experiment with such a knife for cutting curdled milk or some other such very yielding material; indeed, in such material wax is more suitable than steel for knowing the differences dependent on angles less or more acute, because curds are indifferently cut with a razor and with a dull knife.

Simplicio Why does he say "more suitable" rather than "just as suitable"?

Salviati I believe his point is (though he did not expand on it) that if one really wanted to know the exact differences due to sharpness, one could use a wax knife on curdled milk, and then on soft butter, and then on lard or suet, continuing until the sharpness began to make a felt difference, or until its edge began to be dulled by cutting, whereas no such difference would be reached among such substances using a steel knife. So even though "wax knife" is a phrase unheard before, and seemingly ridiculous, there are nevertheless purposes in science that would be best served by such a device, though none that occur in ordinary life.

Sagredo The text proceeds:

It is needful, therefore, not only to attend to hardness, solidity, or heaviness in bodies of different shapes that must divide and penetrate some material, but also to think of the resistances of materials to be divided and penetrated. But since, in making the experiment on the disputed point, I chose material that does penetrate the water's resistance and that in every shape goes to the bottom, the adversaries cannot charge me with any deficiency. I have indeed proposed a more exact way then theirs, to the extent that I have taken away all other causes of sinking or not sinking and have indeed proposed a more exact way than theirs, to the extent all sink with the sole alteration of one grain of weighting, on the removal of which they return to the surface and float. Hence it is not true (again taking their example) that I proposed to experiment on the efficacy of sharpness in cutting by using materials incapable of cutting; rather, I used material proportioned to our need and not subjected to any other variation than that single one which depends on less or more acute shape.

But let us go on a little further and see how entirely unnecessarily they introduced the consideration that they say must be made in the selection of material accommodated to the making of our experiment. We explain by the example of cutting that, just as sharpness is not sufficient for cutting except when it exists in hard material capable of overcoming the resistance of wood (or whatever else we intend to cut), so the inclination to sink or not sink in water should and can be recognized only in those materials that have [92] power to overcome the opposition of water and conquer its bodily character. Moreover, having said that distinction and selection of one material rather than another is necessary, in which there are to be impressed shapes for cutting or penetrating this or that body, according as the solidity or hardness of those bodies is greater or less, I now add that any such selection and caution would be superfluous and useless if the body to be cut or penetrated had no resistance at all and did not in the least oppose cutting or penetration. If knives had to be used to cut through fog or smoke, paper knives would serve equally well with those of Damascus steel. And thus, water having no resistance whatever to penetration by any solid body, any choice of material is superfluous and unnecessary, and when I said above that it was best to choose a material very similar in heaviness to water, that was not because this was needed to overcome bodily resistance of water, but because it is by heaviness alone that water resists submersion of solid bodies. For so far as any resistance [of water] from its bodily character is concerned, if we attentively consider the matter, we shall find that all solid bodies, not only those that sink to the bottom but also those that float, are alike suitable and apt to help us arrive at knowledge of the truth of our controversy.

Simplicio To say the least, that is put cryptically, when the controversy itself is about two very different things—floating and sinking.

Salviati We can say those are very different, indeed opposite, or we can say they are but two aspects of one and the same thing, as we like. You may see this, Simplicio, if you consider how people might have managed without the word "floating" from the beginning to the present. For suppose we had only the phrases *total sinking* and *partial sinking;* would our language be any the poorer?† We might then be inclined to say that our two phrases named two varieties of the same thing, not two totally different things, as *sinking* and *floating* seem to be. This was, perhaps, what Aristotle meant when he cryptically said over and again that change occurs only in the simultaneous presence of opposites, a notion that becomes clear to ordinary mortals only after much study. You will see shortly that our author does indeed reduce all floating to a kind of sinking, in a certain very clear sense.

Sagredo In this, I believe, he takes his clue more from mathematics than from Aristotle, as for example when the straight line and the circle, which at first glance seem to be utterly different, become indistinguishable if the radius of the circle becomes immense, or when a line becomes ever less distinguishable from a point as its length diminishes indefinitely. At any rate, he continues:

Nor am I frightened off from such conclusions by the experience that may be opposed to me of many diverse woods, corks, galls, and moreover thin slices of every kind of stone and metal—quick by natural heaviness to move toward the earth's center—that are nevertheless impotent, either by shape (as the adversaries think), or by lightness, to break and penetrate the continuation of the parts of water and to fend aside their union, so that these rest afloat and do not enter deeply at all. Nor does the authority of Aristotle move me, who in more places than one affirms the opposite of this that experience shows me.

Simplicio Now I have him on the hip, for Aristotle spoke at the end of *De caelo* on the floating of flat bodies heavier than water.

Sagredo Softly, Simplicio; again you are not listening to our friend's exact words. He did not say here that Aristotle made no such assertion as you point out; indeed, the one you mention is to become a principal part of the controversy. Rather, he said that in several places Aristotle affirmed that stones and metals naturally proceed toward the center if nothing resists that motion, which you cannot deny. He then declared that experience shows that some rest on the surface of water, which nevertheless does not resist them at all. We are about to hear how experience shows this:

Again, therefore, I repeat that there is no solid of such lightness, or of such shape, that placed on water does not divide and penetrate its bodily character. Indeed, if anyone will go back with keener eye and look more closely at thin little flat pieces of wood, he will see them with part of their thickness under the [surrounding] water, rather than just kissing with their lower surface the upper surface of [93] the water, as those must have believed who said that chips do not submerge "because they have not the power to divide the tenacity of the parts of water." For very thin chips of ebony, stone, and metal, when they rest afloat, have not only broken the continuation of water [in a line] but are with their whole thickness beneath its surface, and the more so according as the material is the heavier. Thus a thin plate of lead rests as much more below the surface of the surrounding water as is, at the least, twelve times the thickness of that same plate, while gold sinks down below the water level about twenty times the thickness of the [floated] piece, as I shall explain farther on.

Simplicio Excuse me, Sagredo, but since you say you have performed these experiments I must interrupt to ask whether this is indeed true, for I do not see how it is even possible, and I certainly have never seen any such thing. By this rule a gold coin one-twentieth of an inch thick would float an inch under water! Surely no one has ever seen any such thing happening.

Sagredo Indeed not; such a coin could never be made to float at all, for reasons you have already heard that are as old as Archimedes. Nevertheless, it is possible to hammer gold so thin that a small piece of it may be made to float by placing it with extreme care on water. I have done this, finding the simplest way to be to put the bit of dry gold on a piece of thin paper, which when wet will drift slowly down, leaving the undisturbed water holding up the gold. This will then lie notably below the surface of the rest of the water, though whether by twenty times the thickness of the gold I cannot say, the gold being so extremely thin as to be difficult to measure. But nothing floats an inch below the surface of water, or even one-quarter of an inch, and our friend did not assert that anything does; he simply said that the floating body heavier than water always has its entire body lower than the surrounding water, which accordingly it must have had power to thrust aside and penetrate within; and that is all that was necessary here to refute the contrary assertion of his adversaries. Next he says:

Now let us go on to make clear how the water yields and allows itself to be penetrated by even the lightest solid, and bit by bit we shall demonstrate how even from materials that do not submerge one could come to know that shape acts not at all in the going or not going to the bottom, inasmuch as water permits every shape alike to penetrate.

Make a cone or pyramid of cypress, or fir, or wood of similar heaviness (or indeed of pure wax), having considerably height, say four inches or more, and put it in water [in a narrow container, keeping it] base down. At once it will be seen that this will penetrate the water, nor will it be one whit impeded by the breadth of its base. But it will not go entirely under water, sticking out toward the apex. Already it is evident from this that this solid does not rest from going deeper through impotence to divide the continuity of water, having already divided that with its wide part, which in the adversaries' opinion is the least suitable for dividing water. The pyramid having stopped thus, note what part of it shall be submerged, and then turn it point down. You will see that it does not penetrate the water more than it did before, and if you will mark the place up to which it is immersed, anyone expert in geometry will be able to measure [the volumes of] those parts that remain outside the water, which in the one experiment and the other are equal to a hair. Manifestly, from this one may gather that sharp shape, which seemed best suited to fend and penetrate the water, did not fend or penetrate it a bit more than broad and spacious shape.

Anyone who wants an easier experiment may make, of the same material as before, two cylinders, one long and thin and the other short and very wide, putting them in water not sideways but straight up and down. If with diligence you measure the parts of both, you will see that in either of them the submerged part has the same ratio to the part remaining outside the water, and that no [94] greater part is submerged of that long and thin one than of the other, fatter and broader—though the latter is supported on a broad area of water, and the former on one very narrow. Therefore difference of shape introduces neither ease nor difficulty in fending and penetrating the continuity of water, and consequently this cannot be the cause of sinking or not sinking to the bottom [as the adversaries reason]. Likewise, the nullity of operation of difference of shape in [a body's] rising from the bottom of water toward its surface is perceived by taking wax and mixing into it many lead filings until it becomes appreciably heavier than water; make a ball of this and put it at the bottom of water. If you now attach to it just enough cork, or other very light material, to raise it and draw it toward the surface, and afterward mold the same wax into a thin plate or any other shape, the same cork will serve to lift it in exactly the same way.

Simplicio This last experiment is more to my liking than those before, in which mathematicians are needed to calculate volumes. If it is borne out in practice, I cannot think how those who argued that shape alters matters could explain it. What did they say, Salviati?

Salviati So far as I can recall they simply ignored it, for though they were very intent to find instances that forced our friend to give them explanations, he did not ask them in return for explanations of his examples that went against their position. I remember that they ceased to volunteer any explanations after they found that they could not even agree among themselves on some that they proffered at the beginning of the dispute. He told me later that in this controversy ignorance had been his best teacher, because in the search for simple experiments that would vanquish objections by his ignorant opponents, which he would not have needed otherwise, he found out much he had not previously suspected.† In particular, that is how he came to notice one surprising fact that will particularly astonish Simplicio—that a needle floats entirely under water, so to speak, and not just partly beneath the water level as does a wooden ball.

Simplicio What, then, can a needle actually be floated? That is news to me, for Aristotle said explicitly that a needle would not float.

Sagredo Nothing is easier than to float a needle, using that bit of paper I mentioned for the case of gold. As to Aristotle's assertion, that will be discussed later on, near the end of this book. Here it continues:

My adversaries are not silenced by this, but say that what I have said thus far matters little to them; they are content that in one particular case, in material and of shapes suitable to them (that is, a chip and a ball of ebony), they have shown that the latter placed in water goes to the bottom, and the former remains afloat. The material being the same, and the two bodies differing in nothing but shape, they believe they have completely demonstrated and made palpable all they needed to, and have finally carried out their intent. Nevertheless, I think and believe that I can show that their [one] experiment proves nothing against my conclusion.

Simplicio I protest against this notion that a science can be founded on some events, leaving out others, which is an abuse of the method of induction. Aristotle held that a single contradictory experience could overthrow any amount of reasoning, and our friend should not deny that as a profound principle of natural philosophy.

Salviati Far from denying it, I have heard him repeat it so often that it now bores me—or would, if it were not so exactly correct. But Simplicio, it is not always plain and evident that such-and-such a particular experience *does* contradict a reasoned position. You have heard Sagredo say that tiny air bubbles might be present and go unnoticed in certain experiments, leaving them inconclusive, and you have just heard that our friend noticed something about the very experiment adduced against him by his adversaries that they had not noticed at all—and indeed I believe no one had ever noticed it before—namely, that some things may "float entirely under water," so to speak. So once again we are obliged to hear our friend out before we declare his proposed science to stand on faulty foundations.

Sagredo And what if he were to deny the facts alleged by his opponents concerning the one experiment on which they were content to rest their entire case, Simplicio? It was proper that you asked me earlier whether our friend's experiments were borne out by my tests of them, and it is now no less proper for us to be just as severe with this one experiment offered by his adversaries. Since that was his next move, let us go on:

And first, it is false that the ball goes to the bottom and the chip does not; for the chip also goes there whenever there is done with each shape exactly what the wording of our question imports; that is, that both be placed *in* water.

The exact wording was added in the second edition, as follows:

The words were these: That the adversaries holding the opinion that shape affects solid bodies concerning their sinking or not sinking, and rising or not rising, in the same medium, as for example *in* the same water, in such a way that for example a solid that being of spherical shape would go the the bottom, [whereas] reduced to another shape it would not, I, holding the contrary, affirmed that a solid body which, reduced to spherical or any other shape, drops to the bottom, drops there also in any other shape, etc.

We now resume the text of the original edition:

But to be *in* water means to be situated within water, and, by [95] Aristotle's own definition of "place," to be situated [in] means to be surrounded by the surface of the ambient body; therefore the two bodies will be *in* water [only] when a surface of water embraces and surrounds them. But when the adversaries show the little ebony chip not sinking to the bottom, they place it not *in* the water, but *on* the water, where, retained by a certain impediment (which will be explained later), it rests partly surrounded by water and partly by air—which is contrary to our agreement that the bodies must be in *water,* not partly in water and partly in air.

To this he added in the second edition, lest readers should think this a mere quibble:

This can also be made clear from the question raised, which concerned no less those things that sink to the bottom than those that rise from the bottom to float. And who does not see that things placed on the bottom must be surrounded by water?

Sagredo This helps us-see still more clearly our friend's reasons for the precise title of his book, stressing things *on* or *in* water, as I hinted before we started our reading yesterday but could not then completely explain to you. As for those bits of metal or ebony that float under water, so to speak, it would be hard without Aristotle's definition to say whether they are on water or in it, being below the surface and yet not surrounded by water.

Simplicio I see that the matter is not quite as simple as it first appeared, and I approve our friend's having recourse to Aristotle to define the case as well as possible. Yet it is not entirely clear how one can say that a body divides and penetrates water though it remains on or atop it—that is, floats, and yet does not extend above the surrounding surface.

Salviati That is because we usually think of the surface of water not as deformed, but as smooth and flat, so that "on the surface" of water seems to mean atop a smooth surface like that of a table. The ebony chip rests on a watery surface, but on one somewhat below the general surrounding surface, which latter has therefore been divided by it. However we look at it, then, there are problems here of language rather than of fact, and what one peson might consider a mere quibble, others see as an effort to make precise some things not ordinarily noticed, let alone exactly described. In the present case there is also a kind of problem that frequently arises in litigation, where the exact meaning of a contract may come into dispute because of this or that word or phrase.

Sagredo Exactly; and if I may go on reading, you will soon see how, having defended his position on the litigious question, our friend proceeds to deal at length with new facts that at first he did not acknowledge to be relevant to the original debate.

Next note that the ebony chip and ball being placed inside the water, both go to the bottom, but the ball does so more swiftly and the chip more slowly, and the more so according as it is broader and thinner. The cause of this slowness truly is breadth of figure, but these slices that slowly descend are those same that, placed lightly on the water, float. Therefore, if what the adversaries affirm were true, one and the same shape would be the cause, in one and the same water, now of rest and again of slowness of motion, which is impossible; for every particular shape that sinks to the bottom must have some determinate slowness natural and proper to it,† with which it moves, so that every other slowness, greater or smaller, is inappropriate to its nature. If therefore a slice, say four inches square, sinks naturally with six degrees of slowness, it is impossible for it to sink with ten or twenty [degrees] unless some new impediment affects it; and much less could it by reason of that same shape stay still and remain completely impeded from moving. It is necessary that whenever it stops, some impediment intervenes other than just breadth of figure. Something other than shape, then, is what stays the ebony [chip] atop water, the only effect of shape being retardation of motion by which the chip sinks more slowly than the ball.

By the best reasoning, then, it is said that the true and sole cause of the ebony's going to the bottom is its excess of heaviness over [96] the heaviness of the water; the cause of greater or less slowness is broader or more compact shape; but the cause of stopping can in no way be said to be the quality of shape. This is so both because, slowness being increased by expanding the shape, there exists no immense expansion for which there cannot be found a corresponding immense slowness without [ever] reducing the motion to nullity; and [also] because the shapes produced by the adversaries to effect rest [afloat] are the same shapes that also go to the bottom.

I shall not omit another reason, founded also on experience, and (if I am not mistaken) quite conclusive concerning the introduction of breadth of figure and resistance of water against being divided, which have no part in the effect of sinking, or rising, or stopping in water. Take some wood or other material, a ball of which rises from the bottom of water to the surface more slowly than an ebony ball of the same size goes to the bottom, so that clearly the ebony ball more readily divides the water in sinking than that other ball does in ascending. Let this ball be, for example, of walnut wood. Then make a chip of walnut similar and equal to the adversaries' ebony chip, which remains afloat; and if it be true that this rests because of the shape, which is impotent by reason of its breadth to fend the corporeality of the water, then doubtless the walnut chip ought to rest on the bottom when placed there, as less able [than the ball], by this impediment of shape, to divide the same resistance of the water. But if we see and discover by experience that not only the slice of walnut, but any other shape of it will come to float, as we certainly shall discover and see, then let the adversaries kindly cease to attribute the floating of ebony chips to their shape; because the resistance of water is the same upward and downward, and the force of walnut to rise to float is less than the force of ebony to go to the bottom.

Moreover, I may say that if we consider gold in relation to water, we shall find that the former exceeds the latter twenty times in heaviness; hence the force and impetus with which a gold ball goes to the bottom is very great. On the other hand, there is no lack of materials, such as plain wax and some woods, that do not yield even

2 percent to water in heaviness, wherefore their rising in it is very slow and a thousand times weaker than the impetus of gold in [97] sinking. Yet a thin leaf of gold floats without sinking to the bottom, and on the contrary you cannot make a leaf of wax or of one of those woods, which, placed at the bottom, will stay there and not rise. Now, if shape could prevent division and impede descent with the great impetus of gold, how could it not suffice to prevent the same division in ascent by that other material, which has hardly the thousandth part of the impetus gold has in sinking?

It is therefore necessary that what holds the thin leaf of gold or the chip of ebony on water is something that is lacking in those other leaves and chips of materials less heavy than water which, placed on the bottom and set free, rise up to the surface without any hindrance whatever. But they do not lack flat and broad shape. Therefore it is not spacious shape that detains the gold and the ebony afloat. What, then, shall we say it is?

Salviati I think we should notice that although the word *cause* is not mentioned here, cause is what now comes into question. Our friend has destroyed the contention of his adversaries that resistance to division on the part of water can account for the floating of materials specifically heavier than water, showing that if any such resistance existed it would have to be very much weaker against upward than downward motion, contradictory to common sense. Now, if what holds the chip up is not resistance of the medium, he inquires what it may be. In his view this should be something that is always present when the effect is seen, and in the absence of which the effect does not occur. The effect seen in this case is not just floating, but the floating of something specifically heavier than water. Hence the cause sought, while it must not contradict the true cause already ascribed by our friend to every case of floating, must be distinguishable therefrom in observations of floating things heavier than water. In the light of our friend's earlier remark, it will suffice to find an immediate cause, which must be something present when such bodies float, but absent when they sink.

Simplicio It will then remain to discover a cause of this presence or absence, if any such thing can be found in the first place.

Sagredo I believe Salviati's comment was offered not to deny that, but to point out that any such further inquiry may be indefinitely postponed in our friend's science, advances there being made one small step at a time. So here is how our friend proceeds:

For my part, I shall say it is the contrary of that which was cause of sinking to the bottom, inasmuch as sinking to the bottom and remaining afloat are contrary effects, and of contrary effects the causes must be contrary. Without doubt the cause of going to the bottom for the ebony chip or the gold leaf, when they do go there, is heaviness greater than that of water. It is therefore necessary that when they stop [afloat], the cause of this is lightness, which in this case, perhaps through some event not previously observed, comes to be attached to the same chip, rendering it no longer heavier than water, as it was before when it sank, but instead less heavy. Now such new lightness cannot depend on shape, both because shapes do not add or take away weight, and because there is no change of shape in the chip when it sinks to the bottom from that which it had when it floated.

Now take once more the thin leaf of gold or silver, or the ebony chip, and place it lightly on the water so that it rests there without sinking, and carefully observe what effect it has. First you will see how unsound is the saying of Aristotle and of the adversaries that it rests afloat through impotence to fend and penetrate the resistance of the corporeality of water, because it will clearly appear not only that the said leaf *has* penetrated the water,† but that it is noticeably lower than the surface, which sticks up all around the chip and makes a kind of ridge around it within whose depth it rests, swimming. [98] And, according as it shall be of material two, four, ten, or twenty times as heavy as water, its surface necessarily rests below the general surface of the water that many times the thickness of the chip, as we shall prove more particularly below. Meanwhile, for better understanding of what I am saying, look at the accompanying diagram in which the surface of the water extends along lines FL and DB, on which, if there be lightly placed a slice of material heavier than water, so that it does not submerge, this will not at all remain above, but will rather enter with its whole thickness into the water and will drop even farther. This is seen as the slice AI–OI, whose whole thickness goes down in the water, little ridges of water LA and DO remaining around it, while the surface of the water remains noticeably above the surface of the slice. See now whether it is true that the lamina fails to sink because its shape is unable to fend [aside] the corporeality of the water! But if . . .

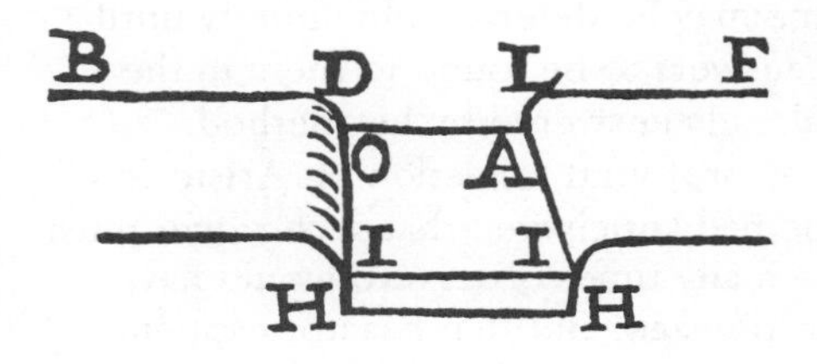

Simplicio Excuse me for interrupting, Sagredo. I take it that you do not pause here because you have already told us that in your experimenting you found it true that the bit of material heavier than water comes to rest below the surrounding water level. I was then content with that, but now for another reason it seems to me impossible. Water being heavier and stronger than air, it should not remain thus indented but should flood over the floating chip.† Therefore, before proceeding, it is necessary to give the cause for the water's holding back and forming this ridge.

Sagredo Now you may see the reasons for Salviati's earlier comment. If our friend were obliged, in his science, to proceed as you say, no doubt in turn you would require him to stop again when he had obliged you, asking him for another cause, and his science would be blocked by an infinite regress, against Aristotle's frequent rejection of such procedures. Having in mind the customary procedure in natural philosophy up to this time, you demand a cause for the formation of water ridges at this point, whereas for our friend's science it suffices that they exist, are seen, and are always present when things heavier than water float, though not otherwise. In short, some questions may be deferred in his science for later consideration, and some may be deferred indefinitely until more is known, permitting answers to be found to them in the same ways as to questions already answered by his method.

You may be sure that if, by profound reflection on Aristotle's principles of physics, anyone had anticipated that such ridges must arise, that fact (which I have many times observed) would have been causally accounted for long ago. But that has not happened, and perhaps no one previously ever looked closely enough to note this phenomenon, unanticipated by reasoning from principles. I think further that if any philosopher had been asked whether such ridges could exist, he would simply have denied the possibility on principles such as those that now make you demand a cause. Now, our friend observed the phenomenon that he will now proceed to discuss at length without immediately attempting to find its cause. When he examines that, later—for this is not one of the questions he postpones indefinitely, but only for the present and in order to get on with his science—you will see how he contents himself with an associated phenomenon and does not propose any new principle that would then have to be carried back to what has gone before. We may then discuss anything we think may be a better explanation, but there would be little profit in doing so before we have heard more about these little watery ridges. And since I believe you will not declare anything to be impossible that is actually observed in nature, but will instead open your mind to a possible incorrectness or at least incompleteness of those physical principles that failed to anticipate and even seem to contradict this observation, I must beg you to defer your demand for a cause of this phenomenon to a more suitable time. Our friend now asks a question similar to yours, but very differently directed, as you will see:

But if the chip has already penetrated and vanquished the continuation of the water [level], and is by its nature heavier than water, what is the reason that it does not go on down, but stops suspended within that little cavity that it has made by its weight in the water? I reply: Because in [gradually] submerging until its [upper] surface arrives at the water level, it loses part of its weight, and it goes on then losing the rest of it in going deeper down and lowering itself beyond the surface of the water, while that makes a ridge and a bank around it. It suffers this loss [of weight] by drawing down and making descend with it the air above itself by adherent contact, which air follows in to fill the cavity surrounded by the little ridges of water. Thus what in this case sinks into and becomes situated in water is not just the gold [here the printer put *ferro* for *oro*] lamina or ebony chip, but a composition of ebony-[or gold]-and-air, from which results a volume [*solido*] no longer heavier than water, as was the simple ebony or the simple gold.

Salviati Allow me to interrupt here (though, as I recall, there comes next a fuller explanation of the author's conception) to say something postponed during our first conversation. The conception, Simplicio, is that though all we *see* floating is that solid body of gold or ebony, that is not all that *is* floating, in a certain sense of the word. For we do not see that slice of air, adjacent to the chip, that has been pulled along with it into an unnatural place for air; that is, a place below the surface of water. Our friend does not demand that we stop using the word *floating* in speaking of the chip, for he likes us to speak in the usual way, but he does ask us to remember that in these cases of the floating of things specifically heavier than water, a certain volume (called by him a "solid") that exceeds the volume of the chip is the true floating object in the author's sense. From this remark you can see why, in the title of his book, he took care to speak of "bodies that stay atop water" and not simply of "floating bodies," for in every sense the chip is a body that stays atop water, but in his sense the chip is not the floating body; rather, it is one *part* of a floating body that consists of the whole chip and some air, without which air the chip would not stay atop water.

Simplicio I do see your point, and though it is certainly an unusual way of talking, we philosophers know very well that sometimes it is necessary to make use of ordinary words in special ways that may perplex some who read what we write. The case here is a bit different from that "wax knife" or "paper knife" a while back, in which the conception is instantly understood though not previously encountered, since ordinarily we have no use for wax or paper knives though we would know how to make them. Here the "ridges of water" defy normal experience and baffle us, since ordinarily water is held back only by hard ridges and does not form any by itself, holding itself back, so to speak. Yet the philosophical mind will think of waves, formed by water and yet in a way holding water back in order to form them. What remains puzzling here is that these ridges remain stationary, a condition not seen in waves.

Yet though I understand our friend's need to describe an unusual situation in unusual phrases, I cannot say I approve his use of the word *solid* to describe a body partly of air, where *bulk* or *volume* would certainly have been better. Neither do I like his saying that the descending chip "draws" adherent air along with it, for air is by nature intractable, and one might better say here simply that it acts to prevent a void, abhorred by nature. I have also other objections, but this time I shall profit from my past overhasty judgments and remain silent until our friend has had a chance to explain his whole conception.

Sagredo Thank you, Simplicio, for this caution. I was about to raise similar and still further objections to our friend's rather cryptic statement, but since I too might have occasion later to retract them, I shall instead proceed with the text.

Salviati Allow me to speak once more before you continue. Your remarks, Simplicio, are very sage, and I know that our friend is still uncomfortable about some expressions that he chose in this matter—not because he feels that any were unjustified, but because they might easily be misconstrued though he could find no others that would be safer from misunderstanding in this difficult business. One that you mention, however, is easy to explain; this is the word *solid.* You think of a solid only in the physical sense, as doubtless others will, but to our friend, as a mathematician, it means precisely what you yourself say—that is, *volume.* For among mathematicians *solid* is commonly used in place of *volume of a solid,* just as *base* was used earlier for *area of the base,* and so on. And thus among inevitable problems of words and understanding we have also this one of technical terms.

Sagredo I am glad you mentioned this, for the word immediately appears again, thus:

And if we attentively consider what and how much is the solid that in this experience enters in the water and opposes its heaviness, you will see that this consists of all that is found beneath the [general] surface of the water, being an aggregate and composition of an ebony chip and an almost equal amount of air, or a bulk composed of a lead lamina and ten or twelve times as much air. But, gentlemen my adversaries, in our question one seeks identity of material, and it is permitted only to alter the shape; therefore remove that air [if you object to it] which, joined with the chip, makes it into a different body, less heavy than water, and do put simple ebony in the water—for certainly you will then see the chip sink to the bottom; or, if that does not happen, you will have won the day. Now, to separate the air from the ebony, nothing more is needed than to wet the surface of this chip thinly with the same water, for water being thus interposed between the chip and the air, that other water around it will run [together] without hindrance and will receive within itself simple ebony alone, as it should do [by the wording of our question]. [99]

But I hear some of the adversaries cleverly turn me against myself, telling me that they do not at all want the chip wetted, because the weight of water thus added, by making it heavier than it was, then draws it to the bottom, whereas to add new weight goes against our agreement, since the material was to remain the same.

To this I respond in the first place that we are dealing with the action of shape in solids placed in water, whence no one should demand them to be placed in water without being wetted; nor am I asking for anything to be done with the chip that is not done with the ball. Besides it is false that the chip would go to the bottom by virtue of new weight added by water, simply by thinly wetting it, since I can put ten or twenty little drops of water on the same chip when it is sustained on water, and those drops, provided that they do not join with the other surrounding water, will not weigh it down so that it sinks. But if I take the chip out and wipe away all water adhering to it, and wet its [whole] surface with just one small drop of water, putting it then back on the water, it will doubtless submerge, the rest of the water [now] running to cover it, the air above not retaining it after this interposition of the very thin veil of water that removes the air's contiguity with the ebony. Without resistance, the air separates itself and does not oppose at all the progress of the other water. Or rather, to say this better, the chip will freely descend because it finds itself already completely surrounded by and covered with water just as soon as its upper surface, veiled with water, arrives at the general level of the water.

Then to say that water can increase weight in things situated in it [100] is most false, because water in water has no heaviness at all, since it does not sink there. Rather, if we shall well consider what is done by any immense bulk of water lying above some heavy body that is situated in it, we shall find by experience that this even greatly diminishes the weight, and that we can raise a very large stone under water that we cannot lift at all when it is removed from the water. Let no one answer that, although water above does not increase heaviness in things under water, it nevertheless increases this on those that swim and are partly in air. This is seen for example by a basin of copper that, when devoid of water and containing nothing but air, will stay afloat, but when water is poured into it, will become so heavy that it will sink to the bottom because of the new weight added to it. To this I again reply, as before, that it is not the heaviness of water contained in the vessel that draws it down [under water], but the copper's own specific heaviness, greater than that of water. For if the vessel were of material less heavy than water, the whole ocean would not suffice to sink it.

Permit me to repeat, as foundation and chief point in the present matter, that the air contained in the vessel before water was poured in was what kept it afloat, inasmuch as a composition is made of air-and-copper that is less heavy than as much water, and the place occupied by the vessel in water when it floats is not equal to that of the copper alone, but to the copper and the air inside that fills that part of the vessel that is beneath the water level. When water is then poured in, air is removed, and there is created a composition of copper-and-water specifically heavier than simple water; and this not by virtue of the infused water having greater specific weight than the other water, but by the heaviness of copper itself and the ousting of air. Now, he would speak falsely who said: "Copper, which by its nature sinks, when shaped as a vase acquires from that shape the power to stand in water without sinking," because copper, however shaped, always goes to the bottom—provided that what you put in water is simple copper. It is not the shape of the vessel that makes copper float, but its not being simple copper that is placed in water, that being an aggregate of copper-and-air. Just so, it is no more and no less false that a thin leaf of copper or ebony [101] floats by virtue of its wide and flat shape, though it is true that that will rest without submerging because what has been put in water is not mere copper, or simple ebony, but an aggregate of copper-and-air, or ebony-and-air. And that is not against my conclusion when, having a thousand times seen metal vessels and leaves of various materials float by virtue of the air conjoined with them, I affirm that shape is not the cause of the going or not going to the bottom of solids placed in water.

And more; I shall not withhold from telling the adversaries that their new idea of not wanting the surface of the chip to be wetted may strike outsiders as a poverty of defenses for their side, considering that at the beginning of our debate no trouble was made about such wetting, and nothing was even said about it. For the origin of the dispute was over the floating of cakes of ice, and it would have been simpleminded indeed to have contended that those must have dry surfaces. Besides which, wet or dry, cakes of ice always float—and in the opinion of the adversaries, by reason of shape.

It may occur to someone to say that wetting the upper surface of the ebony chip should drive it below, though unable by itself to fend and penetrate water—if not by the weight of water added thus, at least by that desire and inclination of the upper parts of the water [its ridges] to rejoin and reunite, from which movement of those parts of the water this chip becomes in a certain sense driven below. That feeble refuge is removed if one considers that, however great may be the inclination of the upper parts of water to reunite, equally great is the repugnance of the parts below to disunite, and the upper parts cannot reunite without driving down the chip, which cannot go down without disuniting the part of water underneath; whence it does not follow as a necessary consequence that for the same respects it should descend. Besides, what may be said of the upper parts of the water can with equal reason be said of the lower—that is, that desiring to reunite, they drive the same chip upward.

Perhaps some of those gentlemen who dissent from me marvel that I affirm that the contiguous air above has the power to sustain [102] that lamina of copper or silver that keeps itself on water, as if I would in a certain way give to the air some kind of magnetic virtue of sustaining heavy bodies to which it is contiguous. To meet all the difficulties as well as I am able, I have been thinking of demonstrating with some other sensate experience how that little bit of air contiguous and above truly supports those solids that, though by nature fitted to sink to the bottom, do no submerge when placed lightly on water if they have not first been wetted completely . . .

Sagredo Here I interrupt my reading to ask Salviati if he can tell us how this "magnetic virtue" came suddenly to be introduced, so unlike our friend's customary hypotheses. Since he was always dead set against attributing virtues or qualities to bodies just to explain some observed unusual behavior in them, saying that that merely multiplied words without any addition to knowledge, I suppose that he here alludes to something that actually occurred during the disputes before he wrote his book; but I cannot imagine what it was that would induce him to break so old a rule.

Salviati You are quite right, and it has already resulted in so much criticism that he told me he would now be sorry he had ever undertaken to defend this notion had it not resulted in his discovering the very striking and curious effect that he is here about to describe. Since he did not wish to withhold from his readers an interesting experiment, he put in his book the remark that had led to it, though (as you say) without sufficient explanation to satisfy readers not acquainted with the story.

It happened that on one occasion when our friend was exhibiting certain of these phenomena at the Tuscan court, he showed that a ball of weighted wax just slightly heavier than water could be made to float by placing it very carefully and slowly in water until just a tiny circle at the top remained unwetted. If then touched, it would sink to the bottom by its own weight. In such cases, he remarked, even a sphere may form one of those cavities with watery ridges, containing a bit of air between it and the general water surface, as in the case of the flat chip illustrated earlier. This bit of air, he said, succeeds in holding the ball afloat. A principal gentleman of the court, Don Giovanni de' Medici, then asked him, "Do you mean that the air has a kind of magnetic power by which it holds up the ball that would not otherwise float?" Our friend did not wish to offend the gentleman, so after a moment's thought he replied, "Why yes, in a sense that is a very apt analogy," meaning only that the bit of air adhering to the wax sufficed to prevent its descent by reducing the specific weight of the entire composition to that of water, as explained before, while the adherence of air was analogous to the adherence of a bit of iron to a magnet. But having given this reply in a public gathering, without explaining it further, he was besieged by challenges in other places to defend this "magnetic virtue." If you now go on, you will see what this in turn led him to discover, after which I shall have a bit more to say on the subject.

Sagredo Your account also explains why he did not name the gentleman, who might have been offended to see his name associated with so controversial a speculation. Well, he proceeds:

> . . . I have found that if such a body has sunk to the bottom, then, without touching it but merely by sending to it a little air that joins with its top, this is sufficient not merely to sustain it, as before, but [even] to lift it and to bring it back up, where (in the same way as before) it stops and rests so long as the assistance of the air joined with it is not lacking to it. For this effect I made a ball of wax, and with bits of lead made this heavy enough so that it slowly descended to the bottom, and I also gave it a very smooth and clean surface. This, placed very gently in water, almost completely submerges with just a bit of the summit exposed, which, so long as it shall be joined with the air, will hold the ball up; but if we remove contiguity with the air by wetting it, it will sink to the bottom and remain there. Now, to make it return to the top and stop close to it by virtue of the same air that before sustained it, push into the water an inverted tumbler, with its mouth down, which will carry with it the contained air, and move this to the ball. Press this glass down until through its transparency it is seen that the air contained in it arrives at the summit of the ball; then slowly draw the tumbler back up and you will see the ball rise, and [it will] come to rest again at the surface if you separate the tumbler from the water with great care, so that it does not stir and agitate the water too much.†

Sagredo For Simplicio's sake I shall remark here that the experiment succeeds very well, and that nothing I know resembles more greatly the drawing of a heavy steel ball along a table by a magnet that is never permitted to touch it. Nor does anything touch the steel ball but air, and of course the table beneath. Just so, nothing but air touches the wax ball that is drawn up in this way. I have even been able to draw up a needle under water, in this same way, by carrying air to it. Our friend now continues:

There exists, therefore, between air and other bodies, a certain affinity that holds them united, so that they are not separated without some little force. The same is likewise seen in water, for if we dip some body into it so that it is completely wetted and then draw it slowly out again, we shall see water follow it and rise up noticeably above the surface before it separates from the body. Also solid bodies, if they are very similar in surfaces so that they fit very exactly together, and no air remains between them that can be drawn apart in their separation and made to yield until the surrounding medium follows to fill the space, will hold most firmly [103] conjoined and are not separated without great force. But because air, water, and other fluids very readily shape themselves to their contact with solid bodies, so that their surfaces exquisitely adapt to those of solids without anything remaining between them, it is in fluids that the effect of this coupling and adhesion is most manifestly and frequently recognized, rather than in hard bodies, which rarely join congruently. This, then, is that "magnetic virtue" which with firm coupling joins all bodies that touch without any interposition of yielding fluids. And who knows but that such contact, when it is most exact, is not a sufficient cause of the union and continuity of parts in physical body?

Salviati This completes the digression on affinity in this book, introduced to justify the concept of magnetic attraction by air that was not our author's phrase or suggestion, but that of another person, which was then seized upon by certain adversaries. Some who have seen the very striking experiment of drawing up a wax ball through water remain just as sarcastic as before about our friend's use of this to illustrate the nobleman's analogy to magnetic power, though no adversary has offered a different, let alone a better, explanation than this universal clinging together of things exactly fitted together, whether of solids or of fluids shaped to them. Everyone knows that when a body is wet it takes some little effort to remove all the water from it, and that it may likewise take some little effort to wet a perfectly dry body all over. Universally known facts are all that our friend alluded to by the word *affinity,* and so far as he is concerned no further explanation is possible when an effect is universal. He is accustomed to say that reason is not required where the senses reach, and I am curious to know what Simplicio may say about such a philosophy.

Simplicio Philosophy? Why, that is no philosophy at all, nor is it any proper kind of science; for science is knowledge of reasoned facts, while all that the senses reach are mere brute facts. Those do not give us causes, which are the goal of true science. I am surprised to hear that our friend, who in these pages has often discoursed of causes, should so take leave of—I was about to say, "his senses," but that would indeed be here a paradox, since it appears that his senses are all that he now means to hold on to. I mean, of course, that I am surprised that he should so take leave of reason as to say such a thing.

Salviati I think that no one who ponders what he has written in this book will seriously accuse our friend of having abandoned reason. No, what he says is that reason has its uses, and they are chiefly of value in organizing sensate experiences. It is an abuse of reason, he says, to apply it on the one hand to universal effects, or on the other hand to inner essences,† and he views the Peripatetic philosophy as engaged mainly in just those two vain enterprises. As to causes, he holds that we gain knowledge when we find that two kinds of phenomena, such as those of speed and those of weight, are connected in the way set forth in his two principles at the beginning of this book. But when we find them universally connected we gain no further knowledge by ascribing hidden causes to speed, weight, and their connection. Therefore he assumes as principles the equality of moments between equal bodies moved with equal speeds, and the equalization of moments by inverse proportionality of speeds and weights, us you have heard. To seek the cause of a principle, he believes, is a vain expansion of empty words. As we saw earlier, he calls *cause* anything that is always present when some effect occurs, and in the absence of which that effect fails to occur. In that view, a universal effect must necessarily lack any cause accessible to us, either to our senses or to our reason, and reason is only abused in the pretended discovery of such causes.

Simplicio Why "necessarily"?

Salviati Because in no way could a universal cause be removed to see whether or not a particular effect would cease.

Simplicio I need hardly say why no philosopher can approve this. Our friend's only philosophical idea here is that of "affinity" between things in exact contact, which he mistakenly thinks may be a sufficient cause of the holding together of natural bodies. Does he not see that all bodies would then be of the same strength?

Salviati No indeed. The very different strengths of steel and wax would be clearly accounted for in terms of exactness of contact of their parts, as we must imagine those parts. Since our senses do not reach so far, that, to our friend, exemplifies a proper field for reason to explore.

Simplicio These ideas are new to me, resembling none of any philosopher I can think of offhand, and I need time to ponder them. I think they may be related to philosophy as shipbuilding is related to voyage and discovery—neither a part of it, nor yet no part of it. For the present I can only jest and say that our friend's novel idea of science can be neither the effect nor the cause of philosophy, and lacks affinity with it.

Sagredo Perhaps, when you have reflected further, you will feel less resistance than before to the notion of separating our friend's new science from philosophy, to the mutual benefit of both. But, however that may be, I see we have arrived at a natural place to end today's discussion, for after the digression we last read our friend takes up more thoroughly the discussion of that supposed resistance of fluids to division that he rejects.

Salviati Then let us follow our custom of adjourning to see the sights of Venice, if Simplicio agrees.

Simplicio If you will excuse me this evening, I wish to ponder quietly these new ideas.

Sagredo Then perhaps Salviati would like to visit Murano while it is yet daylight and see our celebrated glassblowers at work, something of little interest to philosophers.

Salviati I should like that very much, as perhaps there I shall be able to meet Signor Girolamo Magagnati, whom our friend has been trying to entice to Florence to serve Cosimo by expanding glass manufacture there. He has told me that Signor Magagnati is no less accomplished as a poet than as a worker of miracles in colored glass.

Sagredo Not only that, but, like our friend, he is a great fancier of fine foods and wines. With luck, we shall dine with him tonight and you will then remember Venice for its delicacies as well as its beauties. Are you sure you will not join us, Simplicio?

Simplicio Thank you again, but I must decline. I will return tomorrow to resume our accustomed reading and discussions.

THE THIRD DAY

Sagredo As you will recall, we were reading last of the tenacity with which fluids cling to solids, and with which solids themselves cling together when they fit perfectly and have no yielding fluid between them. Our friend turns next to the question whether parts of fluids likewise cling together and thus offer resistance to anything that would divide and separate them, as philosophers have supposed to be the case. As to this, he writes:

Now, carrying out my purpose, I say that we need not have recourse to tenacity that parts of water have between themselves and by which they oppose and resist division, splitting apart, and separation, because no such coherence and repugnance to division exists. For if it did, it would exist no less in the internal parts [of water] than in those nearer to the upper surface, so that the same chip, meeting always with the same opposition and stubbornness, would no less stop in the middle of water than near the surface, which is false. Besides, what resistance can we place in the continuation of water when we see it to be impossible to find any body whatever, of any material, shape, and size, which put into water rests, impeded by the holding together of the parts of this water, so that it moves neither up nor down in accord with the cause of movement it carries? And what greater experience is there of this [presence or absence of resistance] that we seek than what we see every day in muddy water? Placed in a drinking glass, this becomes opalescent after making deposits for some hours, and finally around the fourth or sixth day it becomes pure and limpid, everything having been deposited. Nor can "resistance to penetration" stop those impalpable and insensible atoms of mud that, with their very minimal force, take six days to sink through the space of one foot.

Here, in the second edition, was inserted this paragraph:

Nor let anyone say that a very clear argument of the resistance of water to being divided is our seeing that such tiny little bodies *do* take six days to drop through so short a distance. For that is not an opposition to division, but a slowing of a motion, and it would be simpleminded to say that a thing opposes division which nevertheless allows itself so [readily] to be divided. [104] Nor is it enough for the adversaries to introduce causes retarding motion, [when for them] something is necessary that completely nullifies it and brings things to rest. Anyone who wants to show water's repugnance to divison needs to find bodies that stop in water—not merely move slowly in it.

Salviati I recall the first evening of the debate, where it seems to me that our friend put even more clearly his point that resistance to motion is one thing, and resistance to speed of motion quite another. Professor di Grazia tried to show that water resists division by remarking that, though a sword cuts easily through water, striking its flat surface on water may injure the swordsman's wrist. Our friend pointed out the equivocation from simple motion to speed of motion, for no one questions that a sword merely laid on water and released, even flatways, will swiftly sink.

Sagredo I regret his omission here of this, but his instance of muddy water seems to me even more effective, emphasizing as it does the inability of water to resist penetration by particles that are not even visible to us except in large numbers. He now goes on to ask:

> What, then, *is* this corporeality of water, by which it opposes division? By all that is holy, it is hard to understand if (just as I said above), in attempting to reduce some material [to exactly the specific gravity of water] and starting with something so similar in heaviness to water that a very broad plate of it shall remain suspended, as we say, between two waters, we find this impossible, even though we can get it so close to equiponderance [with water] that [a bit of] lead the size of a quarter of a grain of millet, added to the plate that in air weighs four or six pounds, will carry it to the bottom, and [that grain being] taken away, the plate comes to the surface.

Sagredo Here, Simplicio, I suspect that there is some exaggeration in the figures, though the point our friend is making is conspicuously true to me, who have carried out the attempt with wax up to half a pound. This can be brought so near to equality with water in specific weight that it will stay for a very long time motionless in the middle of a tank filled with slightly salt water up to that point, on which pure fresh water has been laid to complete the filling.† A small lead filing added will then carry it very slowly to the bottom, or the removal of a filing the same size will result in its migration to the surface. But though I have spent many hours and even days in the attempt, starting with such a mass of wax, I have never obtained a weight for it such that it would remain at rest in the midst of pure water. Hence I agree when he goes on:

I cannot imagine (if what I said is true, and it is most true) any minimal force and power that is not even less than the opposition of water to being divided and pushed aside, wherefore we necessarily conclude it to be nothing. For if there were any sensible power here, some large plate ought to be capable of being found that is composed of material, equal in heaviness to water, that would not merely stay between two waters, but could not be pushed lower, or raised, without some perceptible force.

Simplicio Without denying the conclusion, which is in terms of sensible phenomena and experimental failures, I think there is a fallacy in the statement that if a weight or force is smaller than any given force, it is necessarily nothing at all.

Salviati Your objection is logical and strikes at the very basis of our friend's conception of science. I shall not argue the point, but shall merely give you a hint from which you may be able to see why our friend and I, and perhaps Sagredo also, consider the objection to be verbal only. Here, I think you resist a certain way of talking rather than reveal a formal fallacy in reasoning. You agree that, since the conclusion is in terms of the sensible, it is all right to say that something incapable of being detected by the senses—say descent at a rate of one inch per year—may be ignored as nonexistent, so that here we could say there is no motion; but you believe that going beyond the senses to the philosophical reality, there is nevertheless motion, whether we can detect it or not. That is one way of talking, and no one in his right mind will object to it, least of all our friend. But there is another way of talking, introduced I believe by Archimedes,† in which mathematical proof that a quantity must become smaller than any other previously assigned quantity, however small, constitutes proof that that quantity becomes vanishingly small in relation to any assigned quantity and may be treated as nonexistent in relation to any purpose whatever. This idea has turned out to be of great value to mathematicians in dealing with problems that are otherwise very perplexing and perhaps insoluble. Our friend has adopted this way of talking, transferring it from mathematics to physics, the fruits of which become evident in discoveries that philosophers who adhere to the other way of talking are forever prevented from making. Once made, such discoveries about nature are not rendered less interesting or valuable by calling the method of discovery hopelessly fallacious. Do you agree, Sagredo?

Sagredo I do, and I also recall how often our friend used to try, with no success whatever, to explain this to certain philosophers, inviting them first to discover as much as possible and only then to worry about putting together consistently all that is known. He now says:

We can likewise gather the same truth from another experiment, showing how water similarly yields to transverse division. If we place in still and stationary water some great bulk that does not sink to the bottom, we can draw it from place to place without any opposition, by means of a single hair; and in shape it may be whatever you like, so that it may cover a large expanse of water, such as a huge beam pulled sideways.

Someone may perhaps oppose me by saying that if the resistance of water to being divided were, as I say, null, then ships would not need so much force from oars or sails to move from place to place in the tranquil sea or in still lakes. To him who offers such objections, I shall respond that water does not oppose or prevent being simply divided, but water does oppose being divided swiftly, and with so much the greater stubbornness as the [desired] speed is greater. The cause of such resistance depends not on corporeality [105] or anything else that absolutely opposes division, but on the fact that the parts of water divided, in giving place to that solid that is moved in it, must also be moved locally, part to the right and part to the left, and part also downward. This requires that we no less move water in front of the ship (or other body running through water) than water behind and following; for, as the ship advances, then to make space sufficient to receive its breadth, the prow must drive to left and right the nearby parts of water, moving them transversely through a space equal to half its width; and as much displacement must be made in water that, following the poop, runs from the outside parts of the ship toward the middle, as successively fills those places that the ship, advancing forward, leaves empty behind it. Now, because all movements are made in time, and the longer ones in greater time, and it being further true that bodies moved within some time are moved by some power through some distance, they will not be moved through the same distance in less time except by greater power. So wider ships are moved more slowly than narrower ones when driven by equal forces, and the same vessel requires more force of wind or oars the more swiftly it must be driven.

Simplicio In this recital of disturbance of the medium by progress of a ship, pushing water away in front and leaving space to be filled behind, I recognize an old theory of continued motion through a medium after release from the mover,† discredited by Aristotle near the end of his *Physica,* here modified by our friend to show the process while the mover continues to act. Perhaps he thinks that without wind or oars, the ship once set in motion would continue forever, by reason of that ancient theory.

Sagredo Not quite for that reason, though he would agree with your conclusion, which it seems to me you offer ad absurdum. Perhaps you will gather his concept of continued motion from what he adds in the second edition:

But it is not true that any great bulk that floats in still water cannot be moved by any minimal force; it is only true that less force moves it more slowly. If the resistance of water to being divided were in any way sensible, it would follow that to some sensible force, the said bulk would remain completely immovable, which does not happen. Rather, I shall say further that when we advance to a more internal contemplation of the nature of water and other fluids, we shall perhaps discover that the constitution of their parts is such that they not only do not oppose division, but that there is nothing in them that must be divided, so that the resistance felt in moving through water is like what we experience in stepping forward through a great crowd of people. There we feel impediment not from any difficulty of dividing, because none of the people composing the crowd is divided,† but only in moving aside persons already separated and not conjoined. So also we feel resistance in shoving a stick into a heap of sand—because no piece of sand has to [106] be cut, but only moved and raised. There are two manners of representing penetration; one in bodies whose parts are continuous—and here division appears necessary. The other is in aggregates of noncontinuous but only contiguous parts—and here there is no need of division, but only of moving.

Now, I am not certain whether water and other fluids must be considered to be of continuous, or merely of contiguous parts. I am indeed rather inclined to believe that they are contiguous (since in nature there is no other way of aggregating than by uniting or by contact of extremities), and by this I am induced to see a great difference between the coupling of parts of a hard body, and again of the same parts when the same body shall be made liquid and fluid. For if I take, for example, a mass of silver, or some other hard and cold metal, in dividing it into two parts I shall feel not only that resistance that is felt in merely moving them, but another resistance, incomparably greater, that depends on that force, whatever it is, that holds them attached together. And so, if we wish to divide again in two the two parts mentioned, and successively again and again, we shall continue to encounter like resistances, though always less and less as the parts to be divided become smaller and smaller. But when finally, making use of the thinnest and sharpest instruments, which are the very tenuous parts of fire, we resolve this [metal] into perhaps its ultimate and least particles, there will no longer remain in it resistance to division; and not only that, but there will no longer remain in it the power of being divided, especially by instruments grosser than the little stings of fire. And what saw or knife, placed in melted silver, will you find to divide anything that has gone through partition by fire? Surely none, because either the whole shall have already been reduced to the finest and ultimate divisions, or else, if there remain parts still capable of further subdivision, they cannot obtain that except from dividers sharper than fire—as would not be a sliver or bar of iron moved through the molten metal.

Simplicio I am enjoying this more as poetry, with these fiery stings snaking through the ties and severing them so that the sharpest razor can then find nothing to cut, than I am accepting it as science, much less as sober philosophy. Read on, but know that all this proves nothing.

Sagredo Our friend takes from the poets those tricks of phrasing that carry the mind incessantly to examples from sensible experience. He believes that science cannot suffer from this, which serves us as a constant safeguard against our wandering too far from the sensible world into a mere world on paper. He continues:

Of similar arrangement and position I deem to be the parts of water and of other fluids; that is, incapable by their tenuity of being divided, or, if not entirely indivisible, at least surely not divisible by a board or any other solid body manageable by our hands, a saw having to be thinner than the solid to be sawed. Therefore solid bodies put in water do not divide but merely move the parts of water, those being already divided down to minima; and since they are very small and capable of being moved many together, they promptly give place to any tiny corpuscle that descends through them. For however small and light this may be, descending through air and arriving at the surface of water it finds particles of water still [107] smaller and of still less resistance to being moved and pushed than its own pressing and pushing force, whence it dips in and moves that portion of them that is commensurate with its power.

Therefore there is in water no resistance at all to being divided, and there are not even any parts in it that must be divided. I next add that if indeed there were found some minimal resistance (which is absolutely most false), [as] perhaps in trying to move with a hair some great floating structure, or in trying to make a large plate of material equal in heaviness to water sink to the bottom by adding a minimal grain of lead, or by removing one [grain] to make it rise to the surface (which likewise will not happen unless this is done dextrously), note that any such resistance is a very different thing from what the adversaries adduce as the cause of floating of lead leaves and ebony chips. For an ebony chip can be made that floats when placed on water, and even the addition of a hundred grains of lead placed upon it are insufficient to submerge it; yet, wetted, not only will it sink without that lead, but some cork or other light material attached to it does not suffice to keep it from going clear to the bottom.

Simplicio Hold a minute; I am all at sea. In what way are these two resistances "very different things," and how can it be that a hundred lead filings will never sink a precariously floating chip of ebony?

Salviati Your trouble, Simplicio, is partly the fault of the author, who here has taken it for granted that his readers understand the conduct of experiments in these matters. The difference of which he speaks is quantitatively demonstrated, though it would imply a qualitative difference if the effects described were actual, which he denies. He wants us to suppose, contrary to asserted fact, that in some very delicate experiment a certain resistance to division may be found in water. That this must be extremely small, if it exists at all, has already been shown. He now asks us to assume nevertheless that we find it, and that a single particle of lead then overcomes this resistance in an experiment that he—and Sagredo here—have attempted in vain along these lines. Then, he says, such a resistance, even if found, will not assist the adversaries, who need a resistance that is not overcome by even a hundred such grains.

Turning now to their experiment, he says that a chip of ebony can be made that will support in addition a hundred grains of lead without going to the bottom, which is also true, and it is all he needs. But in writing this he has spoken as if just any floating ebony chip will support that much additional weight before submerging, which is not true at all, as you have acutely noted. As you will presently see, there is a maximum thickness of any material that can be made to float within those little ridges of water, which ridges cannot exceed a certain height without collapsing and wetting the chip. Since our friend discusses all these matters in due course, they were certainly not unknown to him. But in writing the above passage added to the second edition he neglected to recite all this in detail, some of which is discussed after this place. Instead, he relied on the phrase "there can be made," thinking of this as modifying all that followed, which does not apply indiscriminately to just *any* chip that would float. But it is evident that if a thin ebony chip weighing, say, half an ounce will float, then one of the same surface and weighing but a quarter of an ounce will also float, and will continue to do so when weighted with many small dry particles of lead.

Sagredo To what Salviati has said, Simplicio, I should like to add that the slightest carelessness of expression, or simple assumption of common sense on the part of his readers, opened our friend to attacks and ridicule by alert philosophers, who consider the goal of all studies and endeavors to be not discovery of facts about nature, but an impossible perfection of scientific language. As a result, our friend has long since given up any hope of informing philosophers about nature and has written in plain Italian† for those who wish to learn and who will try to perceive his meaning when he, either inadvertently or because of the subtlety of certain points, has omitted things plain enough to him and to those who carry out the experiments. To resume:

Now see whether, even given that in watery substance there be found some minimal resistance to division, this can have anything to do with the cause that sustains the chip upon water with a resistance one hundred thousand times that which someone may find in the parts of water. And do not tell me that only the surface of water has such resistance, and not its interior parts;† or that really this resistance is found to be very great at the beginning of division, just as it also seems that greater opposition is found in the beginning of motion than in continuing it; for I shall allow the water to be first agitated, and the upper parts to be mixed with those in the middle and lower down, or else you may remove entirely the top parts and work only with others, yet you will see the same effect. Moreover, when a single hair pulls a beam through water, the beam has to divide precisely the uppermost parts [of water] and also has to begin motion; yet it does begin it, and does divide these [parts, which you say are most resistant of all]. And, finally, put the chip in the middle of the water and hold it still there for a while before you let it go; it will immediately commence motion and continue it to the bottom. Moreover, when the chip stops on water it has already not only commenced to move and to divide it, but has entered [into the water] a good distance.

Sagredo That ends the added section. Before I go on, I want to say that some of these arguments appear to me inconclusive. For example, it may be that there is resistance at the surface of water that does not exist beneath, and of a kind not removed by shaking and mixing the parts. For such a difference might exist not in certain parts of the water as such, but in the surface parts alone, which are in contact on one side with water and on the other side with air. Some special resistance of water to division at the surface only, if it exists, would better explain other things I have observed than does that "affinity" or that "magnetic virtue" accepted by our friend.

Simplicio I was about to raise a similar objection to this last conclusion of his that, because no resistance to division exists inside water, none can exist where it meets with a different element. Nor do I like his insistence that a ship, or even a beam, can be moved in still water by pulling it with a single hair. Surely that is not one of the experiments you have tried, Sagredo?

Sagredo It is not, and I doubt that anyone has tried it. On the other hand, when I first read this I did try pulling by a string quite a heavy timber floating in the canal at my door. I chose a still day, pulled very gently, and was astonished at my own success. When finally I broke the string in attempting to impart more speed too suddenly,

I was again surprised to see the timber continue at the same speed, or very little less, for a long way. Hence I think our friend's exaggeration is perhaps less erroneous than the exaggeration we usually make when we assume that only a hawser could move a large ship in still water, or that horizontal motion stops very quickly after we cease to impart it. But to go on, the first edition continued (without the foregoing long discussion):

Accept it, then, as a true and undoubted conclusion, that water has no resistance at all to simple division, and that it is not possible to find any solid body whatever, of any shape you please, that, placed in water, rests [because it is] prevented by corporeality of water [108] from moving up or down according as it shall exceed or be exceeded in weight (though such excess or difference be insensible) by water. Therefore, when we see the ebony chip (or other material heavier than water) held at the confines of water and air without submerging, in investigating the cause of that effect we must have recourse to some other source than breadth of shape impotent to overcome the opposition with which water defies division. For there is no such resistance, and from what does not exist we do not and must not expect any action.

Therefore it remains most true, as said above, that this [floating of ebony chips] happens because what is placed in this way on water is not the same body as that which is placed in the water. For that which is placed in water is the bare ebony chip, which, being heavier than water, goes to the bottom, while that which is placed on water is a composite of ebony and enough air so that together they are less specifically heavy than water and hence do not sink.

Once again let me confirm what I say. Gentlemen adversaries, we already agreed that heaviness of the solid, greater or less than the heaviness of water, is the true and most proper cause of its going or not going to the bottom. Now, if you want to show that besides that cause there is another, which shall be so powerful as to be able to impede and obviate the sinking of those same solids, and you say that this is breadth of shape, you are obliged, whenever you wish to show such an experiment, first to provide sure circumstances that the solid you put in the water is not less specifically heavy than water, because if you do not do this, anyone can rightly say that not shape, but lightness, was the cause of this floating. But I tell you that when you make show of putting the ebony chip in water, you do not in fact put a solid specifically heavier than water there, but one lighter; because in the water there is, besides the ebony, a volume of air united with the chip which is so light that the two things make a composite less heavy than water. So remove the air and put in the water ebony alone, for thus you will put in a solid heavier than water; and if that does not go to the bottom, you will have philosophized well and I badly.

Now that the true cause has been found for the floating of those bodies that otherwise, as heavier than water, should sink to the [109] bottom, it appears to me that for a complete and distinct knowledge of this affair it will be good to proceed demonstratively, discovering those particular events that take place concerning these effects, investigating the ratios that bodies of different shapes and different materials must have to the heaviness of water in order that they may, by virtue of the contiguous air, remain afloat.

Let there be, then, for clearer understanding, the vessel DFNE in which water is contained, and let there be a lamina, or flat slice, whose thickness is contained between the lines IC and OS. Let this be of material heavier than water, so that placed on the water it dips and drops below the level of this water, leaving the little ridges AI and BC, which shall be of the greatest height they can,† so that if the lamina IS should go still lower by any minimal distance whatever, the ridges could not hold together, but driving away the air AICB they would plunge over the surface IC and submerge the lamina. Therefore the height AI, BC, is the greatest depth admitted by the little ridges of water.

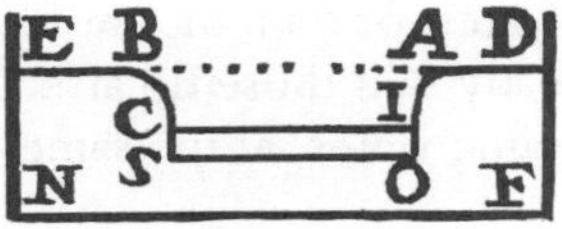

Simplicio Excuse me for interrupting, but what is this height? Surely if such a maximum height exists, it could be measured, and it seems to me that if our friend wants to investigate ratios, as he says, then he should begin by supplying the measures of ratios, which are numbers.

Salviati If you want to know approximately what this height is, for the most part, I can tell you, and our friend spent no little time devising ways to discover this. It is a bit more than one-eighth of an inch. But this is not an easy thing to measure, because the slightest pressure on the chip, as in touching a ruler to it, destroys its floating; nor is it certain, in measuring small distances, to gauge with the eye the height of the surrounding water; yet the ridges must not be touched either, or they will plunge over the chip. Moreover, the maximum height seems to differ slightly with different sources of water, some being purer than others, and with the temperature too. In view of all these things it would not be correct to state one figure for all cases, especially considering that readers have not only other water supplies but other measures of small distances and all the same difficulties in measuring. So our friend prefers to provide information that is universally true, treating of the maximum ridge height in terms of materials obtainable everywhere, as you will see. The slightest difference in thickness of lamina of the same material causes a sensible difference in height of ridge formed, but not in the maximum ridge that will sustain itself, which alone suffices for the ratios our friend will establish. Neither need the source of the water, or its temperature, affect his ratios, though as I have already said those do affect maximum heights. We need only use the same water, at the same temperature, for our comparisons.

Sagredo I might add that if I were told in his book to make a lamina of gold one one-hundredth of an inch thick, I should have to consume a great deal of time doing so, even assuming that the Florentine "inch" meant the same as the Venetian. But by using a lamina of any thickness that will float, I could easily determine what maximum ridge would sustain itself and whether the ridge formed was or was not twenty times as high as the thickness of my lamina. So what Simplicio takes as a failure to be exact, on behalf of anyone wishing to verify our author's findings, is instead the highest degree of exactness possible in communicating freely to everyone what is to be tested, and how that may be done. To continue:

Now, I say that from this [maximum ridge height] and from the ratio that the weight of the material of the lamina shall have to the weight of water, we shall easily be able to determine the greatest thickness in which the said lamina can be made so that it will be sustained by the water. For if the material of the lamina IS shall be, for example, twice as heavy as water, a [floating] lamina of that material will be at most as thick as the maximum height of the ridges [above its top]; that is, as the height AI. Which we shall demonstrate thus:

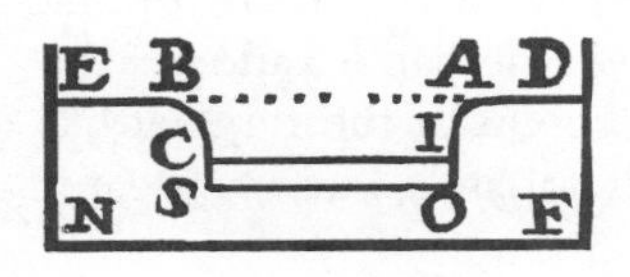

Let solid IS be of double the specific weight of water, and let it be either a prism or a right cylinder, so that it has two plane surfaces, upper and lower, similar and equal and square with the side surfaces; and let its thickness IO be equal to the maximum height of the ridges of water [as defined above]; I say that, placed [with care], it will not submerge in water.

For height AI being equal to height IO, the volume of air ABCI will equal the volume of solid CIOS, and the whole volume AOSB will be double the volume IS. And, inasmuch as the volume of air AC neither increases nor diminishes the weight of volume IS, and solid IS is assumed double in weight to water, therefore as much water as is the submerged volume AOSB, composed of the air AICB and the solid IOSC, weighs exactly as much as this submerged volume AOSB. But when a volume of water as great as the submerged part of a solid weighs as much as the same solid, this sinks no farther, but stops, as was demonstrated by Archimedes (and by me, above); therefore IS will sink no farther but will stop. [110]

And if the solid IS shall weigh three-halves as much as water, it will always float when its thickness is no more than double the maximum height of the ridges; that is, AI. For IS [then] being

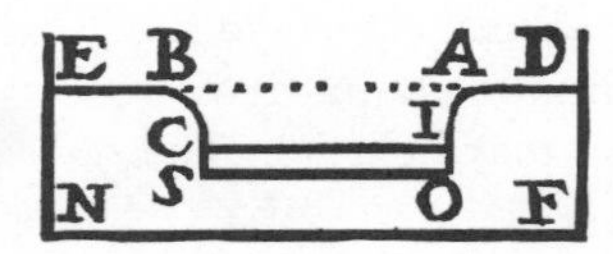

three-halves the weight of water, and height OI being double IA, the submerged solid AOSB will also be three-halves the volume of solid IS; and, since the air AC does not increase or reduce the weight of solid IS, as much water as the volume submerged, AOSB, weighs as much as the submerged bulk; therefore that bulk will stop.

And in sum, generally, whenever the excess of weight of the solid over the weight of water shall have the same ratio to the weight of water that the [said maximum] height of the ridgelet has to the thickness of the solid, that solid will not sink, but any greater thickness [of that material] will sink.

Let the solid IS be heavier than water, and of thickness such that the height of ridge AI has to the thickness IO of the solid that ratio which the excess of weight of the solid IS over the weight of an equal volume of water (to volume IS) has to the weight of the volume of water equal to volume IS; I say that solid IS will not submerge, though if of any greater thickness it would go to the bottom. For as AI to IO, so is the excess of weight of solid IS over the weight of a volume of water equal to volume IS to the weight of the same volume of water; and, compounding [ratios], as AO to OI, so is the weight of solid IS to the weight of a volume of water equal to volume IS; then, inverting, as IO to OA, so is the weight of a volume of water equal to volume IS to the weight of solid IS. But as IO to OA, so is a volume of water IS to a volume of water equal to volume ABSO, and the weight of a volume of water IS to the weight of a volume of water AS; therefore, as the weight of a volume of water equal to the volume IS is to the weight of solid IS, so is the same weight of a volume of water IS to the weight of a volume of water AS. Therefore the weight of solid IS is equal to the weight of a volume of water equal to volume AS; but the weight of solid IS is the same as the weight of solid AS, composed of solid IS and the air ABCI; therefore the whole compound solid AOSB weighs the same as the water that would be contained in the place of this composite AOSB, and hence it will make equilibrium and

[111] rest, nor will this solid IOSC sink farther. But if its thickness IO were increased, the height of the ridge AI would have to increase to maintain the needed proportionality, while, by our assumption, the height of ridge AI is the greatest permitted by the nature of water and of air without the water driving out the air adherent to the surface of the solid IC and filling the space AICB. Therefore a solid thicker than IO and of the same material as solid IS will not remain without submerging, but will sink to the bottom; which was what had to be proved.

In consequence of this that has been demonstrated, many different conclusions can be gathered, by which the truth of my principal proposition becomes ever more and more confirmed, and it is revealed how imperfectly the present question has been philosophized about up to now.

Sagredo You see, Simplicio, that the cause of the ridges need not be determined first. Without that, our friend commences his new demonstrative advance by applying proportionalities.

Simplicio Putting aside all these ratios and proportions, with their compoundings and inversions, until I have had time to study them—if I can find leisure from more profound contemplations in philosophy—I shall say only that I fail to see how the fact of gathering many and varied conclusions from the same thing can more and more confirm a principal proposition, which in this case I take to be that differences in specific weight alone account for floating or rising and sinking in water.

Salviati Excuse me, Simplicio, but again a form of expression has distracted you from our friend's meaning, which is that by deducing new conclusions from an established proportionality, or equality of ratios, we multiply propositions that can be independently put to the test of experiment, permitting new measurements that either confirm or contradict the further proportionalities obtained. When new tests continue to confirm what is gathered, they reinforce our confidence in the original proposition,† showing at the same time that previous investigations, which did not yield what is now gathered, remained imperfect—that is, incomplete. And this is the meaning of the phrase with which our friend began this series of demonstrations; that is, "it will be good to *proceed demonstratively,* discovering those particular events that take place." We have just heard the first stages of a new "demonstrative discovery" as our author conceives that—demonstrative, because proofs are given along the way, and discovery, because the propositions that unfold were not previously known, at least to him. I should be surprised to hear that they were already known to you.

Simplicio Please, Signor Salviati, do not attribute any such implication to my remark; these trifling effects and their mathematical details have nothing to do with philosophy, which is all that I know or can find time to study seriously. If you take this to be science, however, you must also say what happens when new propositions lead to experiments that do *not* confirm them, and thereby contradict the principal proposition, which in this case our friend calls the "true cause of floating."

Salviati Even in such cases, when they arise, our friend's proposed procedure is not without fruit, for discovery of previous error is no less important than discovery of new truth. In the mathematical way of demonstrative discovery, the needed correction often provides information about the source of error itself, which may be of limited extent and may not contradict the original proposition but may rather modify or expand it. An example is the matter we discussed first, when the assumption made by Archimedes was found to apply in most experiments and then was modified to apply to all, our friend adding to his investigations those in which the water level is sensibly changed by immersion of a solid.

Sagredo I think Simplicio will see how new and surprising conclusions may be reached in this way, which are then confirmed by tests that might never have occurred to our friend otherwise, when I go on reading:

And first, it is gathered from the things demonstrated that all materials, even the heaviest, may be sustained on water, not excluding gold itself, heavier than any other body known to us. For considering its weight to be about twenty times that of water, and there being a determinate height of ridges that water can reach without breaking the hold of the adherent air on the surface of the solid that is placed on water, then if we make a lamina of gold so thin as not to exceed in thickness the nineteenth part of the height of the said little ridge, this when placed gently on water will rest there without going to the bottom. And if ebony happens to be eight-sevenths as heavy as water, the maximum thickness that can be given to a chip of ebony so that it may be submerged without sinking will be seven times that of the little ridge [remaining above it]. Tin, eight times as heavy as water, will float whenever the thickness of its lamina does not exceed one-seventh the height of the little ridge.

I do not want to pass over in silence, as a second corollary dependent on the things demonstrated, that breadth of shape is so far from being the cause of floating of heavy bodies that would otherwise be submerged, that even to determine which chips of ebony or lamina of iron or gold may stay afloat does not depend on breadth; rather, that determination must attend only to the thickness of such shapes of ebony or gold, completely excluding from consideration both length and breadth as things that really have no part in this effect.

[112] It has already been made manifest how the cause of floating of the said flakes is merely their reduction to being less heavy than water, thanks to the accompaniment of that air which, together with them, descends and occupies a place in water. If the place occupied, before the surrounding water spreads out to cover it, shall be as large as the water that weighs as much as a given flake, the latter will remain suspended on the water and will not submerge farther. See now on which of the three dimensions of a solid depends the determination of what and how great must be its volume so that the assistance of the air coupled to it may suffice to render it less specifically heavy than water, where it will rest without submerging, and you will doubtless find that length and breadth have nothing to

do in such determinations, but only height or, as we say, thickness. For if you take a flake or chip of ebony, for instance, whose height has the ratio given above to the maximum possible height of the little ridge (which is why it floats), then it does [float]; but not if you increase its thickness even a bit. I say that maintaining this thickness and increasing the area of the chip four or ten times, or diminishing that by dividing it into four or six or twenty or a hundred pieces, it will always remain the same way afloat. But if you increase its thickness by only a hair, it will always sink, though its surface be multiplied hundreds of times.

Now, assuming that to be the cause which, being present, the effect is there, and being removed, the effect is taken away, then, since the effect of going or not going to the bottom is not produced or removed by increasing or diminishing in any way the breadth and length; width or narrowness of surface has no action at all concerning sinking or not. And, given the above ratio of height of ridge to height of the solid, breadth or smallness of surface has no effect whatever. That is manifest from what has been demonstrated above, and from this: that prisms or cylinders having the same base are to one another [in volume] as the heights, whence cylinders or prisms (that is, the chips), be they large or small, provided that they are of equal thickness, have the same ratio to the bounding air that has for its base the surface of the chip and for its height that of the ridge above. Hence volumes are always compounded from that air and that chip which [together] equal in weight a volume of water equal to the volume composed of air-and-chip, for which reason all the said solids alike remain afloat. [113]

In the third place we gather how every kind of shape, of any material heavier than water, may, by benefit of the ridge, be sustained without sinking. Not only that, but some even remain entirely clear of [*sopra*] water and are not wetted except on the lower surface, which touches the water; and such will be all the shapes from whose bases upward the form becomes [always] narrower, as we shall now exemplify by pyramids or cones, which shapes share common properties. Therefore we shall show how it is possible to form a pyramid or cone of any given material that, placed with its base on water, rests without submerging and without wetting more than the base. For explanation of this it is first necessary to demonstrate the ensuing lemma:

Solids whose volumes are inversely proportional to their specific weights are equal in absolute weight.

Let there by the two solids AC and B, and let the volume of AC be to the volume of B as the specific weight of solid B is to the specific weight of solid AC; I say that AC and B are the same in absolute weight. For if the volume of AC were equal to the volume of B, then by assumption the specific weight of B would equal the specific weight of AC, and, being equal in volume and of the same specific weight, one would weigh absolutely the same as the other. But if their volumes shall be unequal, let AC be the larger, and take in it the part C equal in volume to B. Then, since the volumes B and C are equal, the absolute weight of B will have to the absolute weight of C the same ratio that the specific weight of B has to the specific weight of C, or indeed of CA, which is the same specifically. But the ratio of the specific weight of B to that of CA holds, by what is given, between volume AC and volume B, or volume C; and therefore the absolute weight of B is to that of C as volume AC is to volume C. But as volume AC is to volume C, so is the absolute weight of AC to the absolute weight of C. Therefore the absolute weight of B has to the absolute weight of C the same ratio that the absolute weight of AC has to that same absolute weight of C; whence the two solids AC and B have the same absolute weight, which was to be demonstrated.

[114] Having proved this, I say that it is possible to form of any material a pyramid or cone upon any base, which, placed on water, is not submerged, or wetted except on the base. Let the maximum possible height of the ridge be the line DB . . .

Simplicio Excuse me, but I feel some uneasiness already. He says that he will deal with any material whatever in this proof, but then he speaks of *the* maximum possible ridge height. In the first place, I do not see why this, if it exists at all, should be the same for every kind of material, and in the second place I recall his having spoken not long ago about ridges twenty times as high as a gold lamina and only twice as high as an ebony chip. Furthermore, he told us earlier that on a floating ebony chip one might place a hundred granules of lead without sinking it, though surely this would force it lower in the water and thus make the ridge higher. So here it seems to me our author is taking off on a mathematical holiday and exploring ratios that have nothing to do with observed experiments.

Salviati You are mistaken, Simplicio; there is no contradiction here, nor is there any departure from observable events. Your uneasiness stems partly from unfamiliarity with our friend's "demonstrative advance," and partly from inattention to his design of experiments according to the use they must serve. For sometimes, as here, the thing to be shown is not entirely general, as was our author's first proposition about the ratio of volume of water displaced to the volume of the submerged part of the solid. In that first demonstrative advance, the demonstration was equivalent to a theorem in Euclid's sense, whereas in instances like the present the demonstration is equivalent to solution of a problem, in the terminology of Euclid and other mathematicians. For you see that here the proposition states, not that *any* cone of *any* material will float without wetting more than its base, but that it is *possible* with any material to construct a cone that will do so. The demonstration need then show only that if certain conditions are met, say of base angle of the cone, or height of cone with respect to base, the desired result will follow. That suffices to solve the problem how it is possible to construct of any material the desired cone (or, I should say, many possible such cones). Now, to do this, the author needs to make use of the maximum possible height of watery ridge surrounding such a floating object without breaking and flooding its base. Experiment shows that there is such a maximum height for water from a given source and at a given temperature, but (as we said before) it is best to leave this height to be determined by each experimenter, since the water and its temperature will not be the same at every place.

As to varying heights of ridges for different chips, note that our friend has described all ridges in terms of multiples of the thickness of floating lamina. The absolute maximum height must then come out the same regardless of material, analogously to those absolute weights just established in the lemma, and capable of proof in the same way, though our author considered this evident to his mathematical readers and did not offer a separate proof. And finally, with regard to the ebony chip loaded with grains of lead and still afloat, that does not (as you think) bear on the maximum possible height of ridges. For if we can float some ebony chip, it is evident that we can float a thinner one, and one still thinner; hence there will be some chip so thin as to sustain also a hundred lead filings without sinking. But there is an upper limit to the thickness of any ebony chip that can be floated, and such a chip, by floating, gives us the maximum height of ridge to which our author appeals. A chip may float with any lesser height of ridge, and that will be one of the thinner chips just mentioned, which will also sustain some additional weight without sinking. I believe you will now see that this variation in observed heights of ridges for laminae of a given substance does not alter the fixed maximum possible height; it serves only to bring into equality with water the specific weight of the combination of air and material that floats, as our author describes that.

I may add that, if our friend's analysis is incorrect, and if, against his opinion, there *is* a certain small resistance to division by water at its juncture with air, and nowhere else, then that skin resistance, as I may call it, would have exactly the same place in the mathematical analysis of this kind of floating as does this maximum possible height of ridges in our friend's analysis; for such resistance likewise would depend only on the purity of the water and its temperature; it would be the same for all substances floated, and would hold them up to different heights depending on their absolute weight, and so on. Likewise, no such resistance to division could exist *except* at the very surface, or else the same chip would remain similarly suspended wherever we placed it in water, since it was already shown that the water then wetting it, like that lying directly above it, cannot add to its weight in water. As I said earlier, I rather favor this special resistance at the surface, though try as I will I cannot persuade our friend to adopt it.

Sagredo I am glad to hear that I have a companion in this alternative explanation, which also seems preferable to me. For I have noted that, in appealing to Aristotle's definition of place and location, our friend obliged himself to consider as truly *in* water only such things as are entirely environed by water, but he then went on to analyze his chip-plus-air as if it were properly in water—as if it were the entire floating object in these cases and not just the part we see. Yet that floating chip-plus-air is not entirely situated in water, by his own Aristotelian definition, being partly surrounded by water and partly by other air above.

Simplicio Dear me, I should have noticed that egregious error myself; I hope that Professor di Grazia did, for I should like to hear how our friend tried to extricate himself from it. Can you tell me, Salviati?

Salviati First I may say that I am not sure there is an error, despite Sagredo's objection. It seems to me that our friend, in considering "the water surface" as extending straight over the floating chip-plus-air, implied a new definition that saves him from error. The "floating object" touches his "water surface," though the chip alone does not touch it or anything above itself that could be called a "water surface." You may wish to ponder this hint. As to your question, Simplicio, it is a curious thing that di Grazia and the other philosophers were so intent on confirming their own mistaken notion about the role of shape in floating that they looked only for means of supporting that at any cost and did not try to turn our friend's earlier adoption of Aristotle's definition against him here. Also, they valiantly fought to maintain some resistance to division *throughout* water, not noticing that they needed it only at the very surface. Hence no one offered this hypothesis proposed by Sagredo against our friend's. And perhaps it would have contradicted other principles of theirs to oppose him in that way—for example, the principle that the reason of the part must be the same as the reason of the whole.

Simplicio *Eadem est ratio totius et partium;* yes, very true, and this rule is indeed given in Aristotle's *De caelo,* where floating is also spoken of. There was that difficulty for them, and another now occurs to me. This is that physicists must speak of a surface of water, or a surface of air, for there is no physical surface that is somehow between them, what is between them being made of no substance (so to speak) and lying in the province of mathematicians alone. But since not even the surface of water has corporeality, being mere surface and not body, we who wish to follow Aristotle in everything could hardly introduce a bodily quality of the water surface as an explanation against the author. It interests me that you two, his friends and supporters, are willing to attack his conclusions on grounds that we who oppose him are forbidden to use, in the interest of sound philosophy.

Sagredo He does not fear, nor do we, that our friendship or support will be lost by any argument over alternative explanations within his own method. Rather, his method thrives on such debate, since through it things otherwise overlooked may come to light. It seems to him, and to us, that useful science would only be delayed and impeded by polite abstention from controversy, as natural philosophy has long been so impeded by undue concern for the superior authority of metaphysics. But now let us return to the proposition cited above:

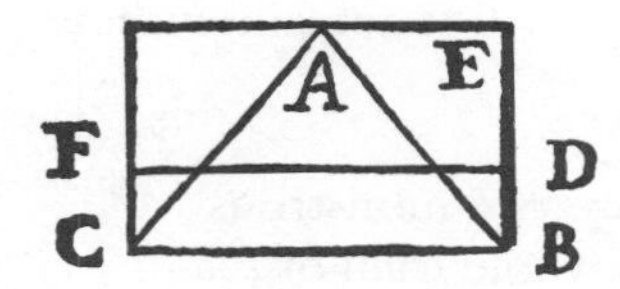

Let the maximum possible height of the ridge be line DB, and let the diameter of the base of the cone, of any given material, be line BC at right angles to DB, and let the ratio of specific weight of the material of this pyramid or cone be in the same ratio to the specific weight of water as the [maximum] height of ridge DB is to one-third the height of the pyramid or cone ABC, whose base has the diameter BC. Then I say

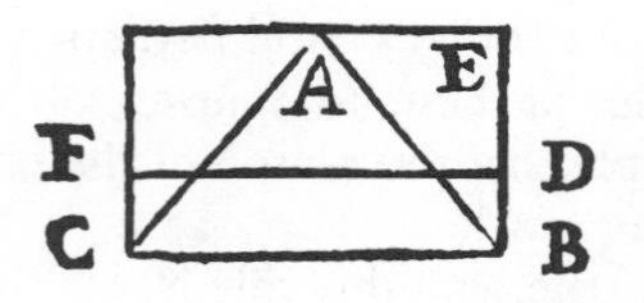

that the cone ABC, and any other lower than it, will rest on the surface of the water BC without being submerged.

Draw DF parallel to BC, and assume the prism or cylinder EC that will be triple the cone ABC. Since cylinder DC has to cylinder CE the same ratio as height DB to height BE, but cylinder CE is to cone ABC as height EB is to one-third the height of the cone, then *ex aequali,* cylinder DC is to cone ABC as DB is to one-third the height BE. But as DB is to one-third of BE, so is the specific weight of cone ABC to the specific weight of water . . .

Simplicio Why?

Sagredo Because DB was taken to be that maximum possible ridge height of which we have spoken, and cone ABC was constructed so as to have the said ratio.

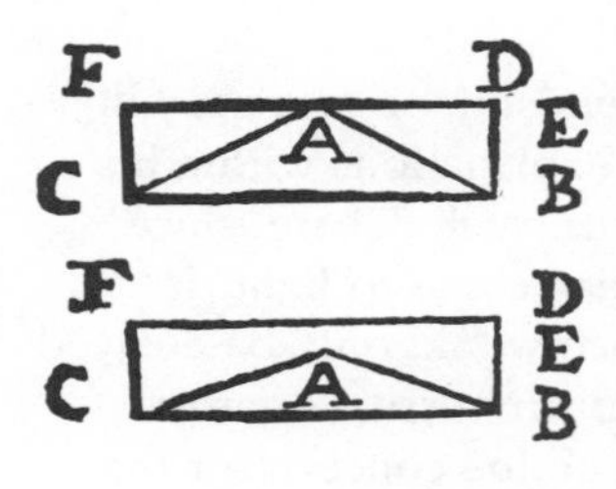

. . . therefore, as the volume of the "solid" DC is to the volume of cone ABC, so is the specific weight of this cone to the specific weight of water; and, by the preceding lemma, cone ABC weighs absolutely the same as a volume of water equal to the volume of DC. But the water that is displaced by the imposition of cone ABC is that which can just fill the place DC and is equal in weight to the cone that displaced it; therefore equilibrium is reached, and the cone will stay without submerging farther. And it is obvious that if we make a cone less high, on the same base, it will also be less heavy, and so much the more will it remain without submerging.

Simplicio I am no mathematician, but to me there are two monstrous contradictions here. Our friend began his whole train of reasoning by asserting, and demonstrating, that the water displaced is always *less* than the solid volume submerged; yet now he wants to have the two precisely *equal.* That is the first monstrosity; have you anything to reply before I state the second?

Salviati Simplicio, you have put your finger squarely on the hopelessness of every possible project of applying pure mathematics to actual physics. It is indeed true that there is no way of reconciling our friend's initial statement, that the displaced volume is always less, with his present assumption that the two are exactly equal. But that can injure only Plato and those mathematicians who vainly aspire to identify pure abstract mathematics with actual physical reality, as when they make the number four the number of fire because pyramids of four sharp points have four sides, and nothing is sharper than fire. It cannot hurt Aristotle, who bans mathematics from physics. Our friend's project is quite different from either Plato's or Aristotle's. He is satisfied if the proportionalities he establishes mathematically are found by the most careful measurements not to be in error by more than a gnat's eyebrow in this matter, or by more than one yard in four miles in the firing of artillery. That imperfection in his science does distress those natural philosophers who want, in addition to knowledge of nature, conclusions that will also be always entirely and forever true, no matter what is measured, or how, in actual events. The actual floating of an ebony chip in a tank of water is much like the launching of a battleship in the Mediterranean. The chip cannot float without displacing one-sixteenth of an inch depth of water covering an area equal to its own, and this will raise the whole water surface one-sixteenth of an inch divided by the ratio of its area to the area of the whole water surface. You may, if you like, calculate the effect, using that first proposition of which you spoke, and apply that correction to the cone's submerged part, and so on. The "monstrous" contradiction in which you have acutely caught our friend will then turn out to be a difference that could not be measured by weighing the displaced water on the most delicate jeweler's balance there is.

Now, logic makes no distinction between the tiniest possible discrepancy and the grossest imaginable, since either one will logically invalidate the most careful reasoning.† Aristotle had the good sense to employ his phrase "for the most part" in order not to be troubled with small or rare departures from his profound conclusions. Our friend could equally well have added the words "except for that difference which may be calculated from my first proposition, here negligible," but he deemed it not worth the space of printing those words for the information of readers who concern themselves with learning new truths about nature, and useless to print them for the benefit of his philosopher-critics, who concern themselves only with universal principles. But if you think he was not aware of what he himself had already discovered and printed at the beginning, or had forgotten his own proof, you are mistaken. Sagredo and I think that his worst fault here was in assuming common sense on the part of all readers—a fault, I must say, that has its counterpart in the writings of his opponents, who assume instead that everyone will gladly reject useful knowledge as long as it is known to fall short of absolute perfection.

Simplicio Your manner of excusing a fault because it is small, or even worse because eliminating it would require a few more words, is most distressing to me. Philosophers, who know that a small mistake at the beginning leads to grievous error in the end, will never acept such slipshod procedures in science, as I am confident the future will show.

But now to my second point. He says that a cone of still less weight will "even more remain without submerging." That is absurd, for there is no greater "not submerging" than not submerging. Seriously, though, by his own reasoning such a cone would not submerge to such a level as would make DB the maximum height of ridge, but only to some higher level. Hence his whole argument would have to be repeated for each and every possible cone of lesser height, which could hardly be done in a lifetime.

Sagredo Right again, as I believe you are also right, Simplicio, about the future: the more our friend's kind of science progresses, the less will philosophers accept it. For, as more and more is learned about nature in this slipshod way, to borrow your phrase, or in this piecemeal way, as I should put it, the more interrelations of things and the more tiny corrections will be discovered,† each of which will give philosophers new grounds for multiplying arguments against accepting as "really" true anything science ever says, where "really" suddenly comes to mean "ideally." My guess is that nevertheless, despite the most determined opposition from future metaphysicians, our friend's method and his projected science will ever gain ground among men of average intelligence who are curious about nature and do not even think it possible to know everything about anything, let alone everything, and who will leave that responsibility gladly, as he does, to philosophers.

Salviati That no science of our friend's kind can ever live in harmony with philosophy is already sufficiently evident in the three years since he first published new discoveries in the heavens—discoveries that you refused even to look at, and that other philosophers are still attempting to argue out of the sky. Now his venture into terrestrial physics is meeting with the same fate, and Professor di Grazia argues that it is insufficient to discuss cones and pyramids, because every possible shape would have to be investigated to disprove the philosophical truth that shape is the reason for the floating of flat bodies, their flat shape making them unable to penetrate the resistance of water to division. This seems to me similar to Simplicio's protest against our friend's having argued a fortiori in the present case, both objections depending ultimately upon a philosophical concept of mathematical certainty different from the concept entertained by mathematicians when they employ that phrase, and even more different from the use of the same phrase by experimenters who are aware of the limitations imposed by units of measurement when applied to continuous magnitudes. Well, since there is no hope of changing the concept of "mathematical certainty" prevailing among philosophers, who neither attempt rigorous mathematical proofs nor soil their hands by attempting careful measurements of brute matter or its actual motion, I will add to Sagredo's prediction one of my own, saying that the day will come when philosophers will recast the very foundations of mathematics, reducing it to a gigantic petitio principii and thus rendering it no longer capable of adding to human knowledge. Yet even after that, I think, mathematicians will continue assisting those who pursue our friend's sciences, receiving in return from them new mathematical ideas, despite the best efforts of philosophers to discourage this slipshod procedure.

Sagredo Well, Salviati, let us admit that Simplicio has demolished our friend's proposition by his requirement of logical perfection and universal scope in every scientific demonstration. Since you and I obstinately consider the proposition new, interesting, and as precise as possible with present means of measurement, we may yet look at the next step in his "demonstrative advance," without insisting that it be called "scientific" by everyone:

It is also manifest how cones and pyramids can be made of any material heavier than water that, placed in water summit or point down, will rest without going to the bottom. For if we take again what was demonstrated above about prisms and cylinders, and if on equal bases to those of the said cylinders we form cones of the same material and three times as high as these cylinders, those will remain afloat, being in weight and bulk equal to the cylinders; and [115] having their bases equal to those of the cylinders, they will leave above [those bases] equal volumes of air contained within the ridges.

This, demonstrated by way of example for prisms, cylinders, cones, and pyramids, could be demonstrated for all other solid shapes; but so great is the multitude and variety of properties and particularities that this would require the writing of a whole volume if we wished to include the special demonstrations for all, and for their sections. But without indefinitely extending this reasoning, I shall be satisfied that any average mind, by what I have explained, can understand how there is no material so heavy, even gold itself, from which it is not possible to form all sorts of shapes that, by virtue of the adherent air above them (and not through resistance of water to penetration), remain sustained so that they do not sink to the bottom. Moreover, to remove a certain error, I shall show how a pyramid or cone placed point down in water will rest without going down, while the same placed base down cannot be made to float—though just the opposite should happen if difficulty in fending the water were what impeded descent, since the same cone is much better adapted to fend and penetrate with its very sharp point than with its broad and spacious base.†

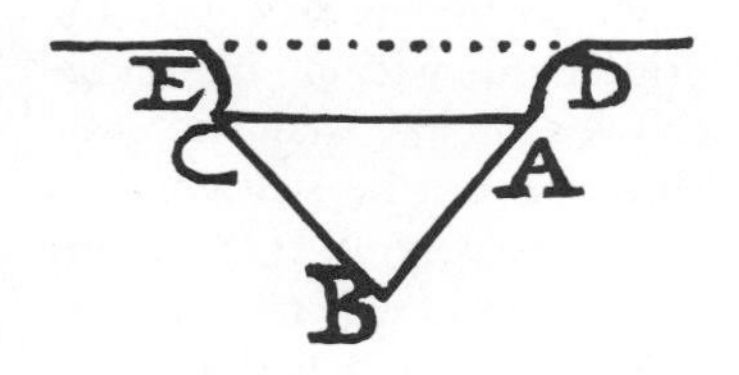

To demonstrate this, let cone ABC be twice as heavy as water while its height is triple the height of the little ridge DACE. I say, first, that put lightly in water point down, it will not go to the bottom. For the aerial cylinder contained within the ridge DACE is equal in volume to the cone ABC, so that the whole bulk of the "solid" composed of the air DACE and the cone ABC will be double that of cone ABC. And since cone ABC is assumed to be of material twice as heavy as water, as much water as the whole volume DABCE, situated beneath the [surrounding] water level, weighs as much as cone ABC, and therefore equilibrium is reached and the cone ABC will not drop down.

I now say, moreover, that the very same cone, placed with the base down, will drop to the bottom, and it is impossible in any way for it to remain afloat. For let there be the cone ABD, twice as

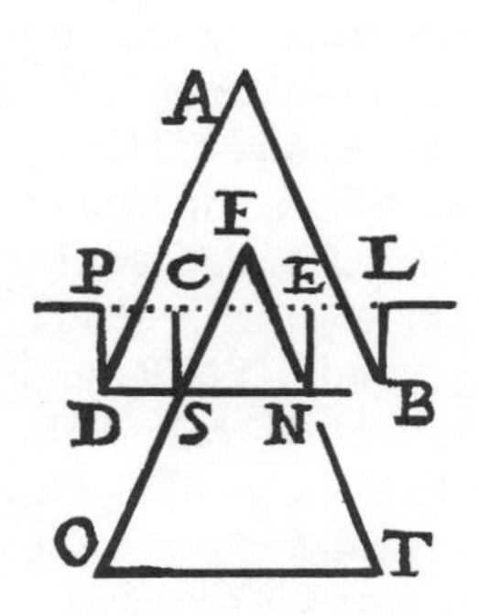

heavy as water, and let its height be triple that of the ridge LB. It is already obvious that this will not rest entirely outside the water, because the cylinder included within ridges LB–DP being equal to cone ABD, and the material of the cone being twice as heavy as water, it is manifest that the weight of this cone will be double the weight of a volume of water equal to cylinder LBDP, whence it will not remain in this state but will descend. I say further that much less will it stop upon a part submerging, which will be understood by comparing with water the part supposed to submerge with the rest that remains outside. Therefore, [116] let the part NTOS of the cone ABD submerge and the point NSF stick out; the height of cone FNS either will be more than half the whole height of cone FTO, or will not be. If it shall be more than half, then cone FNS will be more than half of cylinder ENSC, because the height of cone FNS will be more than three-halves the height of cylinder ENSC, and since we assumed the material of the cone to be specifically twice as heavy as water, the water that would be contained within the ridge ENSC would be absolutely less heavy than the cone FNS, so the cone FNS alone cannot be sustained by the ridge. But the submerged part NTOS, being specifically twice as heavy as water, will tend downward; and therefore the whole cone FTO, both as regards the submerged part and as regards that jutting up, will descend to the bottom. But if the height of the point FNS shall be half the total height of cone FTO, that same height of this cone FNS will be three-halves the height EN, whence [volume] ENSC will be double the cone FNS, and water equal in bulk to the cylinder ENSC would weigh as much as the part FNS of the cone. But since the other, submerged, part NTOS is twice as heavy as water, the volume of water that would compose the cylinder ENSC (and the solid NTOS) will weigh as much less than cone FTO as the weight of a volume of water equal to solid NTOS; therefore the cone will still descend. Indeed, since solid NTOS is [here] seven times cone FNS, of which cylinder ES is the double, the ratio of solid NTOS to cylinder ENSC is as 7 to 2; therefore the whole solid composed of cylinder ENSC and the solid NTOS is much less than double the solid NTOS, whence the solid NTOS alone is much heavier than a volume of water equal to the composite of cylinder ENSC and NTOS. From this it follows that even if the part of the cone FNS were cut off and removed, the remainder NTOS alone would go to the bottom. And the deeper the cone FTO shall go, the more impossible its floating will be, the submerged part NTOS always increasing and the bulk of air contained within the ridge always

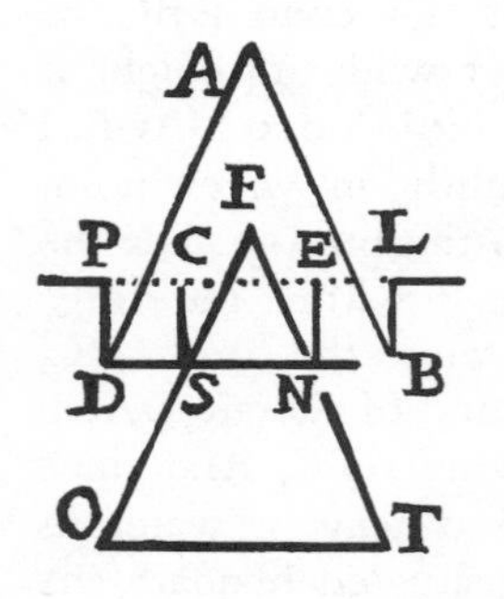

[117]

diminishing, the ridge being always smaller the more the cone submerges.

Such a cone, therefore, which base up and point down is sustained without going to the bottom, must be submerged if placed base down. So those have philosophized far from the truth who have attributed the cause of floating to the resistance of water to being divided as a passive principle, and to breadth of shape that must divide water as the active principle.

Sagredo I may add, Simplicio, that although the author began by specifying a material of double the specific weight of water, and later spoke of a height triple the height of the maximum ridge possible, the curious effect he deduces is not limited to some specific material or size of cone. Those who understand mathematical implications will see at once that the proposition is quite general for cones heavier than water that can be made to float point down, which may consist of various materials and be of various sizes. When I first read this demonstration it seemed to me incredible, though I did remember having seen strawberries dumped in water for washing, of which some floated with the flatter end up. When I tried the same effect with wax cones made slightly heavier than water, I at first placed them most gently on water and released them, and indeed things came out as our author says; but then I became more careless in placing them and found that even when dropped with the point hardly touching the water, they remained afloat, bobbing up and down, whereas no matter how carefully placed base down on water, they went straight to the bottom without delay.

Simplicio Since Aristotle never said that shape prevented sinking in water, I believe you; but I am surprised to hear that base down the cones sank *speedily,* for he does say that shape affects speed of sinking, flat shapes being moved more slowly.

Sagredo I did not observe any sensible difference in speed when two cones were simultaneously released, one point up and the other point down, but I did not try this through any great depth of water or with cones so little heavier than water as to move very gently. You are of course right about Aristotle, to whose precise words our friend tried again and again to induce his adversaries to pay more attention in this matter. Nor did they have any explanation of their own for this palpable experimental contradiction of their position that flatness impedes and sharpness aids sinking in water. Our friend continues:

Fourth, I shall gather and conclude reasons for that which I [next] proposed to the adversaries, that is: That it is possible to form bodies of any shape and any size that, though by their nature they would go to the bottom, may by the aid of the air contained within the ridges remain without submerging.

The truth of this proposition is very evident in all those solid shapes that terminate in a plane surface on the uppermost side; for if we form such figures of some material equal in specific weight with water and then put them in water so that the whole is covered, it is manifest that they will remain stationary in all places, given that such material could be adjusted to a hair in weight to equal water; and in consequence these would remain at the very surface of water without making any little ridge at all. Now, if with respect to material such figures can remain without submerging although deprived of help from a ridge, it is clear that without increasing their volumes, they might be increased in weight by the weight of as much water as would be contained within a ridge made around the upper plane surface, with the aid of which [ridge, if filled with air] they would remain afloat; though wetted they would go to the bottom, having been [increased in weight as just said above and] made heavier than water. In shapes, then, that terminate above in a plane, it is clearly understood that the presence or absence of the little ridge can prevent or permit descent. But in those shapes that become thinner toward the top one may, and with good reason, doubt whether the same may be done, especially with shapes that go on to end in a sharp point, as do cones and thin pyramids. For these, then,

[118] as more in doubt than any others, I shall try to show how they also are subject to the same events of going or not going to the bottom, and they may be of any size you please.

Sagredo Here, before continuing, I should remark that I did not have uniform success in verifying by experiment the things about to be deduced in this demonstrative progress. It was here that in some cases the same cone would now be kept afloat as expected and again would sink despite my diligence, and I discovered, as I remarked earlier, that the presence of tiny bubbles seemed to me responsible for success. Also, I found that when the ridge is of small diameter it is difficult to release a wax cone held by the tip so that all above the bottom of the ridge is kept quite dry. I do not mean to deny that our friend may have overcome these and other difficulties in performing the experiments, but I cannot confirm these as I can the others. Here is his demonstration:

Therefore let there be the cone ABD, made of material specifically as heavy as water; it is manifest that when placed entirely under water, this will rest at any place (I mean whenever it weighs exactly as much as water, which is almost impossible to effect), and if any the least heaviness is added to it it will go to the bottom. But if it will drop downward [initially and] very lightly, I say it will make the little ridge ESTO and that there will remain out of water the point AST, of height triple that of the ridge ES. This is manifest, because the material of the cone weighing equally with water, the submerged part SBDT remains indifferent to motion up or down, and the cone AST being equal in volume to the water that would be contained within the ridgelet ESTO, it will also equal this in weight, and in consequence equilibrium will be reached and the whole will be at rest.

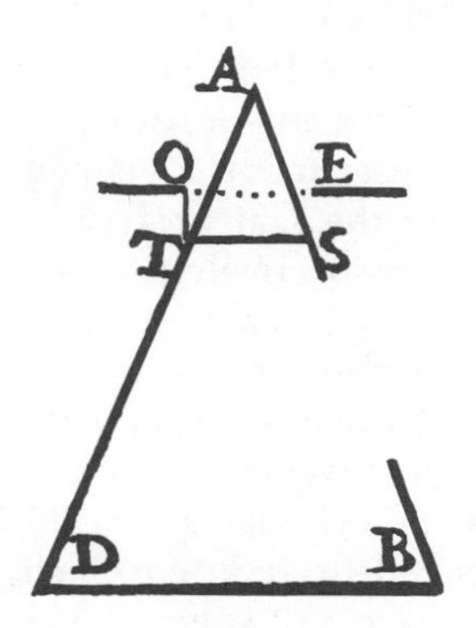

Now the question arises whether the cone ABD can be made

heavier so that when placed [entirely] in water it will go to the bottom, but [still be] not so much heavier as to remove from the ridgelet the faculty of being able to sustain it without submerging. The reason for questioning is this: that if indeed when the cone ABD equals water in specific heaviness, the ridgelet ESTO sustains it, [it should do so] not only when point AST is triple the height of the ridgelet ES, but even more when a smaller part remains outside the water. For although the cone in descending makes the point AST diminish, and the ridgelet ESTO also diminishes, nevertheless the [protruding] point diminishes in greater ratio than the ridgelet, being reduced according to all three dimensions, while the ridge is reduced only in two, the height being [maximum and] constant. Or we might say because the cone ST shrinks in the ratio of the cubes of the lines that successively make up the diameters of the bases of the cones projecting, and the ridgelets are reduced in the ratio of the squares of the same lines, the ratios of the projecting parts are always as the three-halves power of the cylinders contained within the ridgelets. Thus, for example, if the height of the projecting point were double, or equal to, the height of the ridge, in these cases the cylinder contained within the ridge would be much larger than the said projecting point, for it would be three-halves, or triple, [respectively,] so that it might grow in force to sustain the whole cone, since the submerged part does not weigh down at all. Yet when some heaviness is added to the whole volume of the cone, so that even the submerged part is not without some excess of heaviness over the heaviness of water, it is not clear whether the cylinder contained within the ridgelet, as the cone sinks, will be able to reduce to such a ratio with the emergent point and to such excess of volume over its volume as to be able to offset the excess of specific heaviness of the cone over that of water.

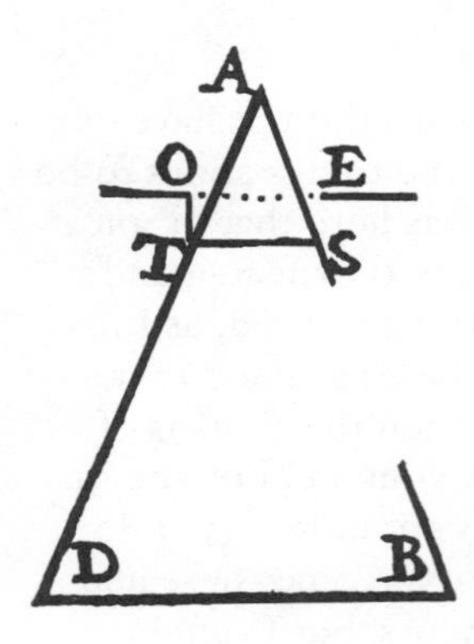

[119]

This doubt arises because, although in the cone's descent the projecting point AST diminishes, whereby the excess of heaviness

of the cone over heaviness of water also diminishes, the fact is that the ridge also narrows and the cylinder contained therein diminishes. Yet it will be demonstrated that the cone ABD being of any size, and made originally of material very similar to water in heaviness, some weight can be added to it by which it may descend to the bottom when placed under water, but may yet, by virtue of the ridgelet, be stopped [at the surface] without submerging.

Sagredo Before going on to the demonstration I shall repeat that precisely this is the hard part in actual experiment; that is, to hold the cone by its very tip just at the position of equilibrium before releasing it, since any little further drop as a result of holding it a bit too high, or any excess wetting as a result of holding it a bit too far down in the water, may destroy the effect. Again I do not deny that the author may have succeeded in consistently overcoming this difficulty by some device or technique that has not occurred to me. But I had little, and equivocal, success, made inconsistent by the occurrence of those tiny air bubbles mentioned before.

Let there then be the cone ABD of any size and of specific heaviness equal to that of water; it is manifest that if placed lightly in water this will rest without submerging, [with] point AST, of height three times the height of ridge ES, projecting. Now imagine the cone ABD to be deeper, so that there projects beyond the water only the apex AIR, half the height of AST, with its ridge CIRN. Since cone AST is to cone AIR as the cube of line ST is to the cube of line IR, but cylinder ESTO is to cylinder CIRN as the square of ST is to the square of IR, cone AST will be the octuple of cone AIR, while cylinder ESTO will be quadruple of cylinder CIRN. But cone AST is equal to cylinder ESTO; therefore cylinder CIRN will be double cone AIR, and the water that would be contained within the ridge CIRN will be double in volume and weight to cone AIR, and therefore the

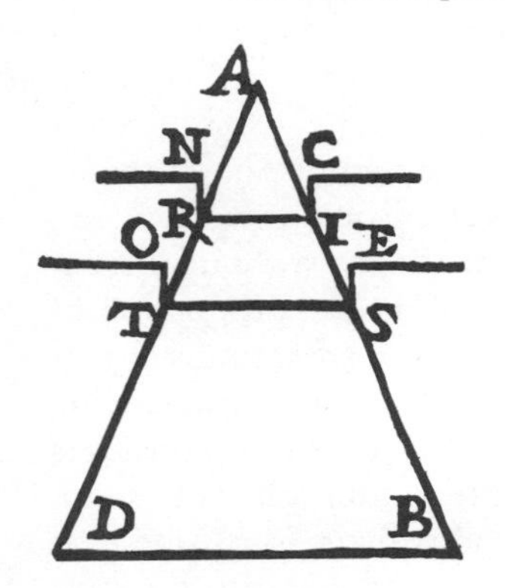

ridge has power to sustain double the weight of cone AIR. Hence, if there shall be added to cone ABD as much weight

[120]

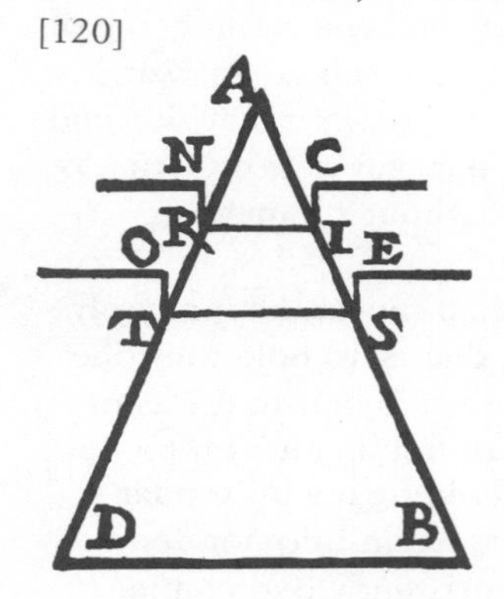

as that of cone AIR, which is one-eighth the weight of cone AST, it can also be well sustained by the little ridge CIRN, without which it would go to the bottom, cone ABD being by this addition of weight equal to one-eighth the weight of the cone AST rendered specifically heavier than water. But if the height of the cone AIR were two-thirds the height of cone AST, that cone would be to cone AIR as 27 to 8, and cylinder ESTO would be to cylinder CIRN as 9 to 4, or as 27 to 12, and therefore cylinder CIRN would be to cone AIR as 12 to 8, and the excess of cylinder CIRN over cone AIR would be to cone AST as 4 to 27; therefore if to cone ABD there shall be added as much heaviness as 4/27 the weight of cone AST, which is a bit more than one-seventh, it will remain afloat and the height of the projecting point will be double the height of the ridgelet. What is thus demonstrated for cones happens exactly in pyramids, no matter how sharp either may be; from which it is concluded that the same event will happen even more easily in other shapes that progressively terminate in less sharp summits, with the aid of more spacious ridges.

All shapes, therefore, of whatever size, may go or not go to the bottom according as their summits are wetted or not wetted; and these events being common to all sorts of shapes without the exception of even one, shape has no part at all in the production of this effect of sometimes going to the bottom and sometimes not, as does their now being joined with the air above and again being separated therefrom. In the end, then, for anyone who considers rightly (or, as people say, looks at this matter with both eyes open), the cause is reduced to, or rather really and truly is, that same true physical and primary cause of floating above or going to the bottom—that is, excess or deficiency of the heaviness of water with respect to the heaviness of that corporeal volume that is placed in the water. For, just as a slice of lead as thick as a knife blade, which by itself alone goes to the bottom when put in water, will stay afloat if there be attached to it a four-inch cork—because then a solid is put in water that is not, as before, heavier than water, but less heavy—so the chip of ebony, by its nature heavier than water and therefore going to the bottom if alone placed in water, if placed upon the water with a wafer of air that goes together with the ebony in descending, and if of such size that together they form a composite less heavy than water in that volume which is submerged beneath the water level, this will not go on but will stop; and for no cause other than that universal and common one—which is that bodies less specifically heavy than water do not go to the bottom. Whence he who should take a plate of lead an inch thick, for example, and four inches square, and should attempt to make that float by imposing it gently, would be wasting his time; for when it has descended a hairbreadth farther than the [maximum] possible height of the ridges of water, it would be covered and would sink. But if, as it was being lowered, someone should make around it some edges that would hold back the spreading of the water over this plate, which edges were heightened enough so that within them there could be held enough water as to weigh as much as the plate, then doubtless it would sink no farther, but would rest sustained by virtue of the air contained within those said edges; and in a word, there would have been formed a vessel with a lead bottom. If the thinness of the lead shall be such that a very small height of edges [less than the maximum water ridges] should suffice to encompass as much air as would keep it afloat, it would come to rest [even] without such edges—but not without that air, for air by itself can make edges sufficient, through a very small height, to restrain the flooding by water. Hence what floats in this case is again a vessel filled with air by virtue of which it remains above without submerging.

[121]

Simplicio I find it interesting that our friend reasons in this way from a vessel with a lead plate for its bottom, with edges of any material added as needed, to a fanciful lead "vessel" having water for its edges. Though I am not sure that his unique cause, taken from Archimedes, is the correct one for these cases, nevertheless I cannot disapprove his having reasoned by analogy from a true vessel to a vessel with watery sides, for Aristotle tells us that analogy is a correct means, and sometimes our only means, for reasoning to causes in natural philosophy. And this also helps me see how, in that long mathematical argument preceding this, one may think of the pointed cone sticking out above the water level as resembling a cone that is supported by a little cylindrical cup, of some substance whose bottom is pierced by the point of the cone, which cup is able to sustain the small excess of weight of the whole cone over the weight of water, most of that cone's weight having been removed by the submerging of its greatest part.

Sagredo Very well put. Our friend has now completed his "demonstrative advance" in this matter, as far as he thought it useful to pursue. He concludes this part of the book with an amusing reversal of roles between himself and his adversaries, pretending that he has undertaken to defend a fallacious position that they then correct by adopting tactics similar to those he used in analyzing the matter of floating chips heavier than water. I believe that Simplicio will find this especially interesting as a kind of polemic argument that has novel aspects. He writes:

In conclusion, I shall try to resolve every difficulty by another experiment, should anyone remain in doubt concerning the operation of this continuity of air with the thin plate that floats, and then I shall put an end to this part of my *Discourse.*

I fancy myself in a dispute with some of my adversaries about whether shape has any action in increasing or diminishing a weight's resistance to being lifted in air. Undertaking to sustain the affirmative, I declare that a bulk of lead in the shape of a ball will be raised with less force than that some lead made into a very thin, broad plate—as something that must, in a broad shape, fend a vast amount of air, but little of it in that more confined and compact shape [of a ball]. Then, to show how my opinion is true, I first suspend the ball by a thin cord and put it in water, tying the cord that holds it to one of the arms of a balance that I hold in the air, while to the other arm [122] I proceed to add weight until that finally lifts the ball out of the water, which requires, say, thirty ounces of weight. Then I alter the same lead into a flat, thin plate, which I similarly place in water, suspended by three cords so as to hold it parallel to the surface of the water, and in the same way I add weights to the balance pan until this plate is lifted and taken from the water, showing thus that thirty-six ounces are insufficient to separate it from the water and lift it through the air. Based on this experiment I affirm that I have fully proved the truth of my position.

Simplicio Let me be clear about this. Since thirty ounces sufficed to lift the same lead when it was in the shape of a ball and was still wet, and six ounces is far more than a reasonable weight for water merely wetting the same volume of lead in a more distended shape, our friend seems to mean that 36 ounces failed to lift the plate when it was still not entirely "taken from the water," but rather was only raised very slightly above the water and was still touching the surface. Is that right?

Sagredo Yes, Simplicio, as shown also by the fact that he did not name the weight (which might be, say, 36½ ounces) that sufficed to separate plate and water entirely. You will see in a moment his purpose in this. Here he continues:

An adversary now steps forward, and by having me tilt my head down makes me see something I had overlooked. He shows me that in [just] leaving the water, the [lead] plate draws in its train another plate, of water, which, before it divides and separates from the lead plate, rises up above the level of the rest of the water more than [the thickness of] a knife blade. He then repeats the experiment with the ball, showing me that only a very small quantity of water adheres to its narrow and compact form. Next he tells me that it is no wonder if, in separating the thin, broad plate from the water, I feel much more resistance than I do in separating the ball, since one must raise a great deal of water along with the plate, which does not happen with the ball. He also calls it to my attention that our debate is whether resistance to being raised [in air] is greater in a broad lead plate than in a ball—not whether a lead plate with a great quantity of water [attached] resists more than a ball with very little water. Thus he shows me that first to put the plate and the ball in water, in order then to test their resistances in air, is beside our point, because we are arguing over the raising in air of things situated in air, and not over the resistance made at the boundary of air and water by things partly in air and partly in water. Finally, he makes it palpable to me that when the thin plate is in air and free from the weight of water, it is raised by the same force, to a hair, as was the ball.

Simplicio Except, he should add, as to the weight of water wetting the broad flat shape as against the water wetting the ball, a difference that must be more than the weight of a hair. These continual leaps he makes to conclusions that are not philosophically or mathematically sound destroy any possible value in his "new science," at least to the mind of a careful logician, which is only the first step in becoming a true philosopher.

Salviati Then you, Simplicio, believe that with the correction for the weight of water wetting a plate as against that wetting the ball, the balance would require exactly the same counterweight in both cases?

Simplicio Certainly; there can be no doubt that one and the same wieght requires one and the same force to lift it, in the same medium.

Salviati I think not, or rather, in the present instance I think that rule will not philosophically and mathematically hold.

Simplicio Well, if you are going to deny the most elemntary principles of mathematics and physics, there is no use my arguing with you; for, as Aristotle says in his *Physics, contra negantes principia non est disputandum.*

Salviati On the contrary, I am speaking *ad pleniorem scientia,* if you want to make this a battle of Peripatetic phrases. For look you: By your own insistence, we must take into account the area of the plate, larger than that of the ball made of the same lead, and therefore requiring more water for its complete wetting. I have to admit that on this you are correct, mathematically and philosophically. But I then point out that air has a certain buoyant force on bodies situated and lifted in it, just as does water, and the volumes of both ball and plate are altered by the layer of water wetting them. Therefore there is still another correction to be taken into account in your philosophically exact science, for the net density of ball-and-water will not be quite the same as the net density of water-and-plate, the two having different proportions of lead and water.

Simplicio Oh, yes; but this difference (if it could be detected) would be trivial.

Salviati It would be trivial in our friend's science, and, like the weight of the wetting water, was neglected there in order to get on with the job of learning useful things about nature, previously unknown. But in your exact philosophical science, if I understood you, it is sinful to neglect anything whatever—which is why you blamed our friend for neglecting the difference in weight of the wetting water.

Sagredo Gentlemen, if I allow you to continue this argument we shall never get on with our job, which is to learn the things said by our friend in this book. It seems to me sufficient to say that he and Simplicio draw the line differently between the important and the trivial, so that your real dispute is over taste in science, not progress in science. Let us get back to the imaginary dispute in our friend's book:

Having seen and understood these things, I can do nothing but admit myself persuaded and thank my friend for his having made me understand what I had previously overlooked. And then, apprised of these phenomena, I reply to my adversaries that our [original] debate was whether an ebony ball and an ebony chip equally go to the bottom in water—not an ebony ball, and an ebony chip that is joined with another chip of air; and, further, that we

[123] were speaking of sinking to the bottom, or not, in water; not of

what happens at the boundary of air and water to bodies partly in air and partly in water. Neither were we dealing with the greater or less force required to separate this or the other body from air. And, finally, I tell them that air resists and (so to speak) weighs [up] just as much in water, as water weighs [down] and resists [motion] upward in air, and that it is the same work for us to push under water an inflated bladder as to lift in air that bladder filled with water. Likewise it is true that the same work is required to push a tumbler or similar vessel down into water when it is filled with air, as to raise it above the surface of water, holding it mouth down while filled with water, which is constrained to follow the tumbler that contains it and to be raised above the rest of the water into the region of air, just as air is forced to follow the same vessel beneath the water boundary—until, in one case, water rising above the lip of the vessel precipitates itself within and drives out the air, and in the other case the lip of the tumbler comes out of the water, and entering into the region of air, the water falls out and air comes in from below to fill the hollow of the vessel.

From this it follows that he no less transgresses the bounds of our debate who produces a chip joined to a quantity of air, to see whether it sinks to the bottom in water, than he who would test resistance to being lifted in air with a lead plate joined together with as much water.

Sagredo That completes the demonstrations of the book, which next turns to discussions more philosophical in character. Our session today has been already long, so let us stop for today and ask Simplicio if he will resume the reading tomorrow, the final parts being more in accordance with his main interests.

Simplicio I shall gladly do that, but before we stop I should like Sagredo to report on these last statements, quite remarkable if true, in which it was said that to push air under water in a glass requires the same work as to lift a glassful of water in the air. Is that really true?

Sagredo It is quite true, with a very small correction that I noticed when I thought it through. But, instead of beginning with that, I think you will be interested to hear how I persuaded myself of the facts. The case our friend described was first of a tumbler in its normal position, pressed down into water until the water reaches its very brim, and then of an inverted tumbler filled under water and lifted, base up, until the brim reaches the surface.† That at once suggested to me the idea of pressing an *inverted* empty tumbler into water, then turning it right side up so that it would fill, and finally lifting it back up into the air. Would the effort of pressing it under water, empty, be the same as the effort of lifting it out again, filled? Until I read our friend's statement, that would have seemed to me a silly question, there appearing to be no connection whatever between the two actions. But the *position* of the glass containing air should not matter, so it seemed that the effort should be the same if our friend was correct here. I then saw at once why it *would* be the same, or very nearly. When an empty glass is forced into water, no water is destroyed to make room for it; the water simply moves aside. But restrained by the sides of the tub or other container, it is forced to rise, and the effort I feel is simply that of raising as much water as was pushed aside—that is, one glassful of water, very nearly. And of course that is the amount of water raised in a glass in the normal way.

Simplicio Not so, my friend, for when you press your inverted glass into water, water rises into the glass, compressing the air within it, as it cannot do in our friend's case, with the glass base downward.

Sagredo Very good, Simplicio, but in your haste to instruct me you did not allow me quite to finish. Before I do, let me ask you how much you think this effect would be, speaking of an ordinary drinking glass pressed into water upside down until just submerged?

Simplicio I suppose the water would rise in it a quarter-inch or so, perhaps more. Surely it would appreciably reduce the volume of air contained.

Sagredo Philosophically speaking, the volume is appreciably reduced, but as a matter of experience I can tell you that the air in the glass is hardly compressed at all when the glass reaches submersion, and your "entrance of water" is more a phrase than an observable fact.

So concede to me that this objection of yours does not destroy the rather wonderful near equivalence of efforts I set out to investigate, inspired by our friend's statement. Now tell me what other objections occur to you.

Simplicio I am so astonished that your conclusion was anywhere nearly correct, and so puzzled by the failure of water with its great weight to overpower and compress the yielding air in the inverted glass—for which I must accept your assurance from experiment—that nothing more occurs to me. Pray proceed.

Sagredo Well, since glass is denser than either air or water, it is evident that its natural downward tendency assists us when we push the tumbler down in water and impedes us when we lift the water in it up into the air. Hence there is a difference between the two efforts, though if the glass is thin the difference is not great, and certainly the felt effort in either action far exceeds that of our lifting an empty glass, even a thick one. Now, if we go back to our friend's actual words, we see that he avoided the extra effect of lifting the *glass* in air by speaking only of the water contained in the glass; but he did not mention the assistance we receive from the weight of the glass when we are forcing its contained air into water.

Simplicio And so, by your own admission, Sagredo, our friend's science is far from exact, as science must be to win even that name among philosophers, let alone to command their respect.

Sagredo Oh, come now, Simplicio; "far from exact" is but rhetoric here. Our friend offered a statement so surprising to you that you asked me to testify to its truth or falsity. I extended it in a way even more unexpected and assured you that a certain inexactness which your science took to be serious was, in fact, hardly observable. Finally, I pointed out another inexactness, the effect of which you would find it very difficult to measure, and now you seize on that to discredit an entire science without which we would have had no inkling of several very interesting phenomena of nature.

Salviati I am on Sagredo's side, Simplicio, for a reason he has not mentioned. Our friend's approach to his science gives it a curious quality that promises always greater and greater exactness—without end, so to speak. For each objection that has been raised in this particular instance suggests something further to be considered, and at the same time shows us how to go about removing the objection. Being only human, I like our friend's promise of continual improvement, by means open to anyone, much more than I like your assurance that perfection in science was already achieved, as far as that is possible, by Aristotle, long before I was born.

Simplicio That idea of gradually approaching toward truth by means that leave us incapable of ever reaching it, particularly when such means are available to anyone, merely fills me with apprehension. But I grant to Sagredo that "far from exact" was not the right phrase to express my objection to this piecemeal way of doing science.

Sagredo No; if you had said "see how far from exact our friend's science is," you would have hit the nail on the head, for that is precisely what we can do at every stage—see *how* far our conclusions depart from the most careful observations we can make.† Just to know that is no small achievement. And now I want to add a conjecture of mine about the way our friend probably hit on this conclusion about moving air in water and water in air, which seemed so surprising and yet was adduced by him without any proof or even any explanation. Surely he was not trying to mystify his readers, so it must have seemed to him that anyone could see why the two cases amount to no more than lifting the same amount of water in two different ways—one when it is spread out over a large surface, and the other when it is collected in a glass. The key to this, I believe, is the importance our friend habitually attaches to *measuring* things—the very habit that makes it possible, in his science, for us always to be able to judge how nearly we have approached toward an unattainable perfect exactness.

I say, then, that our friend probably arrived at his conclusion when he was performing that "magnetic virtue" experiment, in which the effort of pushing down the empty inverted glass† to bring air to the submerged object is strongly felt. It would be natural for him to ask himself how that effort could be measured—a question he would have answered in the way I have indicated, and so swiftly, for his long practice in seeking measures, that he may have had his answer before he had even finished bringing the little wax ball to the surface. Every man supposes that whatever he has quickly seen must be easily seen by others; and that, I think, is why our friend did not explain his statement to his readers.

Simplicio Well, I am certainly glad that I detained you for my final question, in answering which you have awakened in me real curiosity about these problems of the behavior of bodies in water, which two days ago seemed to me so trivial that I wondered at all the time and effort our friend lavished on them. Now I begin to see how it comes about that men otherwise intelligent, but not trained in philosophy, spend so much time on matters that no philosopher has ever deigned to consider. If our friend is right in supposing that a science restricted to sensate experiences and necessary demonstrations is possible, then some day that may perhaps develop to a point at which it will even deserve the attention of philosophers.

Salviati Our friend is confident that what he has presented, and is going to present in parts of physics he has studied but not yet published, will come to be the elements of entire new sciences that will occupy men more able than himself in time to come.† But he does not think that mankind will ever prefer those over the pursuit of chimeras through endless mazes, enticed by grandiose promises that everything in the universe may be understood simply by reading philosophy.

Sagredo Let me predict that if philosophers ever do take an interest in our friend's sciences as they develop, they will then claim credit for the very pursuits that they now strive to discredit, arguing that without their critiques those sciences would have remained mistaken and could not have been purified by such clumsy beginners as our friend who ignored metaphysics. They will make even Archimedes into a Platonist, though he left not a single sentence on philosophy.

Simplicio Philosophy takes all knowledge for its province, so that if new sciences provide any kind of *knowledge,* they will prove to have been part of philosophy. But it will be long indeed before our friend's small beginnings can be extended to throw light on the essences of things and the causes of motions, which alone interest natural philosophers. If and when that happens, the true sources of our friend's thought will necessarily be found in the philosophy he studied.

Salviati Let this profound conclusion place a seal on today's discussions, as we embark once again to explore Venice and her wonders.

THE FOURTH DAY

Salviati It was a good idea of yours, Sagredo, to have Simplicio (when he arrives) begin to read the arguments of our friend's adversaries, stubborn Peripatetics as they are. Having seen some of their printed attacks, and projected replies to them by our friend, I shall be interested to hear the learned professor's opinion of interpretations of Aristotle by those self-styled Aristotelians who lack his acuity.

Sagredo We have not long to wait, for here he comes now. Good morning, Simplicio; we eagerly await the conclusion of the book, and even more your opinion of this more physical and less mathematical part of it.

Simplicio Good morning, gentlemen. The arguments we shall hear today are indeed more in my line than were those of yesterday's meeting, and they reveal new aspects of our friend's judicious reasoning. In at least a few places, I see, he has better understood the meaning of Aristotle than have his opponents. To oppose him they even follow interpreters who alter the text of Aristotle, which surprises and distresses me. Philosophy seems to have fallen on evil days at Pisa, perhaps because that university does not take care to foster differences of opinion as I told you has long been done at ours.

Salviati Before you begin today's reading, I may be able to shed some light on our friend's approach. He is deeply interested in the observation of nature, an activity in which he acknowledges Aristotle's preeminence in antiquity, whose teacher, Plato, disdained observation of mundane things as being of dubious value to philosophers, and even dangerously misleading. Accordingly, our friend assumes that Aristotle took care in reporting observations, so that when his texts are uncertain or seem wrong, our best guide to his meaning is what we ourselves find by observation today, and not the verbal ingenuity of commentators who made few studies of nature at first hand. For scientists must assume that nature remains faithful to rules that it is their task to discover in any way they can. Bodies did not float or sink, or indeed fall through the air, any differently in Greece two millennia ago from the way they do in Italy now. So Plato was wrong in supposing that *only* mathematical truths remain forever unchanged, the same being true of at least some physical truths revealed by careful study of sensate experiences. If Aristotle was equally mistaken in excluding mathematics from physics, experience and mathematics together may yield a science more certain in dealing with the real world than is either the philosophy of Plato or that of Aristotle. Our friend expects that science will accord with much, but not with all, that each of them said, so he feels no obligation to support or to contradict philosophers of either school, except in the narrow domain of sensate experience and of necessary demonstrations based on it.

Simplicio I shall keep what you have told me in mind, for it has often seemed to me that our friend implied much more.† Still, such a science as you describe appears to me hopelessly incomplete and very defective, neglecting the most important and profound parts of philosophy. But now let me begin reading the concluding section of his book:

I have now said all that comes to mind as showing the truth of the side I undertook to maintain. It remains for me to consider what Aristotle wrote on these matters at the end of *De caelo,* noting two things. One is that it being true, as demonstrated, that shape has nothing to do with simply being moved or not being moved upward or downward, it appears that Aristotle at the outset of his speculation held the same opinion. This, it seems to me, may be gathered by examining his words, though it is true that in his attempt then to give the reason for this effect (like those who in my opinion have not well found this out) I must examine second how it appears that he came to regard [action of] breadth of shape separately from this operation [of simply moving.]

[124] As to the first matter, here are Aristotle's precise words: "Shapes are not the cause of moving simply up or down, but of moving more swiftly or more slowly, and the cause for which this happens is not difficult to see."

Here I first note that the terms coming into the present consideration are four; that is, motion, rest, slow, and swift; and since Aristotle names shapes as the cause of the slow and the fast, excluding them from being the cause of absolute and simple motion, he necessarily excludes them likewise from being the cause of rest, whence his view may be stated thus: "Shapes are not the cause of absolutely moving or not moving, but of the slow and the swift." For if anyone should say that it was Aristotle's intent to exclude shapes from being the cause of motion, indeed, but not from their being the cause of rest, so that his intent was to remove shapes from being the cause of simply moving, but not from their being the cause of resting, I shall ask whether we must, with Aristotle, understand that all shapes, universally, are somehow the cause of rest in those bodies that would otherwise move, or only some particular [shapes], as for example broad and flat shapes? If all indifferently, then every body will rest, because every body has some shape; and that is false. But if some particular [shapes] alone may somehow be the cause of resting, as for example broad shapes, then the others ought to be somehow a cause of moving; for if from seeing some bodies of compact shape moving that then, expanded into plates, are stopped, I could infer breadth of figure to be [included] separately in the cause of that rest—just as from seeing such plates rest that then, compacted, move, I might with like reason affirm that united and compacted shape has some part in causing motion, since, removed, it impedes this. Then that is directly opposed to what Aristotle says: that is, that shapes are not causes of movement.

Besides, if Aristotle himself had granted (and not excluded) shapes to be causes of not moving in some bodies that, shaped otherwise, would move, he would improperly have proposed questioningly, in the words immediately following: "why it happens that broad and thin plates of iron or of lead stop upon water," since the cause of that would be right at hand, that is, breadth of figure.† Conclude, therefore, that Aristotle's conception in this place was to [125] affirm that shapes are not causes of absolutely moving or not moving, but only of moving swiftly or slowly, which [interpretation] may be the more believed as in fact that opinion is most true.

Now, such being the viewpoint of Aristotle, and this consequently appearing to be at first glance more against the positions of my adversaries than favorable thereto, their interpretation must be not precisely this, but [this is] something [only] partly accepted by some of them, and [is] partly rejected by others. This can easily be understood to be the case, there being an explanation conformable to the sense of celebrated interpreters, which is that the adverb *simply,* or *absolutely,* in the text should not be linked to the verb *to be moved,* but rather to the noun *cause,* so that the sense of Aristotle's words would be to affirm that shapes are not absolutely causes of being or not being moved, but are indeed causes *secundum quid;* that is, in such a way that they are named only as assisting or concomitant causes.

That position is accepted and taken for true by Signor [Francesco] Buonamico in book 5, chapter 28 [of *De motu*], where he writes thus: "There are other, concomitant, causes by which some things float and others submerge, among which the first place is held by the shapes of bodies" and so on. Concerning that position there arise for me various doubts and difficulties, by which it seems to me that Aristotle's words are not capable of such a construction and sense; and the difficulties are these.

First, in the order and arrangment of Aristotle's words, the particle *simply,* or as we say, *absolutely,* is attached to the verb *move* and separated from the word *cause,* which is a strong presumption in my favor, because the text says: "Shapes are not the cause of moving simply up or down, but indeed of slowly or swiftly," and does not say "Shapes are not simply the cause of moving up or down." When the words of a text, transposed, have a different sense from that of their sound in the order in which the author disposed them, they must not be permuted. And who will wish to affirm that Aristotle, wishing to write a proposition, would arrange his words in a way in which they might carry a sense very different and even contrary? Contrary, I say, because understood as they are written, they say that shapes are *not* the cause of moving, while transposed they say that shapes *are* a cause of moving, and so on.

Moreover, if Aristotle's intention had been to say that shapes are not simply causes of moving up or down, but only causes "according to which," it would not have been necessary to add the words [126] "but they are causes of the swifter or slower." Rather, adding this would have been not merely superfluous, but false, inasmuch as the whole feeling of the passage would import this: "Shapes are not the absolute cause of moving up or down, but are indeed the absolute cause of the slow or the swift," which is not true, since the primary causes of the less or more swift were attributed by Aristotle in text 71 of Book 4 of his *Physics* to the greater or less heaviness of moveables as compared with one another, and the greater or less resistance of the mediums dependent on their greater or less corporeality; and those were put by Aristotle as primary causes, and only those two were named in that place, after which shape was considered at text 74 as rather an instrumental cause of the force of heaviness that divides either by shape or by impetus. And truly, shape by itself, without force of heaviness or lightness, does not operate at all.

Simplicio Thus far I have not interrupted my reading, being in good agreement with our friend against the other interpretations. I pause here, indeed, not to object, but to praise the manner in which he introduces that very important passage from Aristotle's *Physics,* concerning ratios of speeds in motion, to support his reading of the last chapter in *De caelo.* This shows his awareness that Aristotle would not define or give rules for the same thing in two different ways, though in two different books. This is a very important consideration in interpreting a doubtful text, for Aristotle was no less consistent throughout his books of natural philosophy than is mathematics, or sensate experience. Our friend continues:

I add that if Aristotle had had the idea that shape was in any way a cause of moving or not moving, then the inquiry that he immediately undertook, in the form of a question why it happens that a plate of lead may float, would have been inappropriate. For if he had just said that shape was in a certain way a cause of moving or not, it would not have been necessary to question the cause why the lead plate floated, next attributing the cause to shape and reasoning in this manner: "Shape is a cause *secundum quid* of not going to the bottom, but now it is doubted for what cause a thin lead plate does not go to the bottom, and the answer is that this comes about by the shape." Such reasoning would be indecent in a child, let alone Aristotle. For where is the reason for doubt? And anyone can see that if Aristotle had deemed that shape were in some way a cause of floating, he would have written (without the questioning form), "Shape is in a certain way the cause of floating, whence the lead plate, thin and broad in shape, floats."

But if we take Aristotle's proposition as I say, and as it is written, and as in fact is true, the train [of text] runs along very well, both in its introduction of the slow and the swift, and in the question, which falls right in place. And the text will speak thus: "Shapes are [127] not causes of moving or not moving simply, up or down, but indeed of moving more swiftly or more slowly; but if that is so, it is questioned what is the cause that it happens that a broad thin plate of iron or of lead floats," and so on. The occasion for this question is immediate, since at first glance it appears that shape is the cause of this floating (since the same lead, or a less amount, of different shape, goes to the bottom), while we had already affirmed that shape has no action in that effect.

Finally, if Aristotle's intention in this place had been to say that shapes—though not absolutely—may at least in some way be a cause of moving or not moving, I place in consideration that he no less mentions upward movement than downward. For in his examples he does not produce any other experience than that of the plate of lead, or the ebony chip—materials that by their natures go to the bottom but that by virtue (the adversaries say) of shape remain afloat. Anyone who can should then produce another experience of materials that by their nature float but are held on the bottom by shape. Since that is impossible to do, we conclude that Aristotle here did not want to attribute any action to shape in simple moving or not moving.

But that he then philosophized correctly in investigating the solutions of the questions he proposed, I do not undertake to sustain. Indeed, various difficulties that present themselves to me give me occasion to doubt whether he has completely explained the true cause of the present business. I shall go along advancing those difficulties, ready to change my opinion whenever it is shown to me that truth is other than what I say, to the admission of which I am more inclined than I am to contradiction.

Aristotle, having proposed the question "whence it comes about that broad plates of iron or lead float," adds (as if strengthening the grounds of questioning this) "inasmuch as other things, smaller and less heavy, if they shall be rotund or long, as would be a needle, go to the bottom." Now, that I question—indeed, I am certain that a needle, placed gently on water, remains afloat no less than those thin plates of iron or lead.

I cannot believe, though it has been reported to me, that anyone would say in defense of Aristotle that he meant a needle put in [water] not lengthwise, but erect and point first; yet to leave not [128] even that refuge, however weak (and one that I think Aristotle himself would reject), I say that the needle should be understood to be laid according to the dimension named by Aristotle, which is length. For if another dimension than that named can and should be understood, I should say that also plates of iron and lead would go to the bottom if anyone were to put them in edgewise and not flat; but since Aristotle says "broad shapes do not go to the bottom," one must understand "placed flat," and so when he says "long shapes, as a needle, though light, do not remain afloat," one must understand "placed lengthwise."

Simplicio All this seems to me very sound, but the philosophers must have objected, for in the second edition he added:

Besides, to say that Aristotle meant a needle put in point first is to make him speak great foolishness, because in this place he says that small particles of lead or iron, if round or long like a needle, do go to the bottom, so that in his view a sliver of iron also cannot remain afloat. And, if he believed that, what simplicity would it not be for him to add that not even a needle, put in erect, would stay? What else is a needle than many such [long, pointed] granules placed one on another? It is too unseemly for such a man to say that a single sliver of iron cannot float, and that a hundred of them put together one on another will not even float.

Simplicio When I read this I was at first puzzled, for what Aristotle next said was that, by reason of their size, dust and metal filings will float, so it appeared that our friend had misread the text. But going on I found what our friend must refer to; for Aristotle does indeed say that the smaller the materials are divided, the more surely they will sink, as things more easily dividing the water. The first edition went on:

Finally, either Aristotle believed that a needle placed lengthwise on water remained afloat, or he believed that it did not so remain. If he believed it did not remain, he was surely also able to say so, as in truth he did; but if he believed and knew that it does float, for what cause did he omit to question, when raising the problem of the floating of broad though heavy shapes, whence it comes about that long thin shapes, though of iron or lead, float? Especially since there seems to be more occasion to question concerning long, thin shapes than flat, thin ones—as Aristotle makes clear by his not having raised this question.

Simplicio And in the second edition we read here:

No less an incongruity would be saddled on Aristotle by one who, to defend him, might say that Aristotle meant a very thick needle, and not a thin one, because then I would ask him what Aristotle believed about a thin needle? He would have to reply that Aristotle believed it would float, whereupon I should accuse Aristotle again of having avoided a more marvelous and difficult problem while introducing one easier and less wonderful.

Simplicio After which the first edition continues:

[129] We shall therefore say freely that Aristotle believed that broad shapes alone floated, but long, thin ones like needles did not, which [belief] is nevertheless false, as it is also false of rotund bodies [that they do not float], since from things demonstrated above [by me] it can be gathered that little globes of iron, and even of lead, [may] float in the same way.

Then he propounds another conclusion that is likewise far from the truth, which is that some things swim in the air by their smallness, such as very minute dust, and thin leaves of beaten gold.†

But to me it seems that experience shows us that not only in air, but even in water, this does not happen. In water, there descend even those particles that muddy it, whose smallness is such that they are not seen except in many hundreds together. Dust, and gold leaf, are not sustained in air either, but descend below, and only go wandering when strong winds raise them or some other agitation of the air moves them—and that also happens in any commotion of water by which deposits are raised from the bottom and muddy it. Aristotle cannot mean this impediment [to sinking] from commotion, of which he makes no mention, nor does he name anything other than the lightness of those minima, and the corporeal resistance of water and air—from which it is seen that he was dealing with air that is quiet and not agitated or moved. But in that case neither gold nor dust, however minute in size they may be, are sustained, but quickly descend.

Simplicio Here, though I will grant that Aristotle perhaps did not look into these matters as closely as he could have done, because they are light and trifling in comparison with the weighty problems that most interested him; yet no less has our friend concluded without considering some cases in contradiction of his position, that can be derived from his own principles. For, just as we can change wax from less heavy than water to more heavy by embedding in it tiny bits of lead, so nature may sometimes endow bits of matter with the same specific heaviness as water, which would then remain in it wherever they were placed, and move about without ever settling or floating. And the same may be true of particles floating in air, made specifically heavy in exact balance with air by the mingling of particles of fire with any other substance.

Salviati Well said, Simplicio; but what you have adduced is, as you see, of no help to Aristotle and still less to our friend's adversaries, for all it demonstrates is the need always to avoid any show of final knowledge in science. On this our friend would agree with you, for he is accustomed to say that there is no event in nature, not even the least that exists, which will ever be completely understood by any theorist. I shall convey to him your criticism, which I know he will gladly have, and I believe he will take more pains in the future to heed it in whatever he may write; for, as he said a while ago, it is more agreeable to him to be corrected than to contradict. What he objects to most strongly on the part of his adversaries in this debate, as in others, is that they carry principles to extreme lengths without regard for ordinary experience. Here he has inadvertently done that himself, so hard is it to found a new kind of science without falling into habits learned from the old. Perhaps in those future times concerning which we dared to speculate not long ago, philosophers will come to restrict themselves to prudent criticisms of the kind you have just raised.

Simplicio You are most kind, and I should like to say that my purpose in those former heated arguments, more often than our friend may think, was to urge similar cautions upon him, and less often to insist on greater loyalty to Aristotle's conclusions than to his methods and his principles. But to proceed:

He next passes on to refute Democritus who, by his [Aristotle's] testimony, would have it that igneous atoms, which ascend continually through water, push and hold up those heavy bodies that are very broad, whereas narrow ones sink down because only a few such atoms oppose and reject them.

Aristotle confutes this position, I say, by asserting that that [support] should occur even more in air [than in water]. Just so, Democritus himself objected against his own doctrine, but then, having raised the objection, he brushed it lightly aside by saying that those corpuscles that ascend in air do not make their impetus unitedly. Here I shall not say that the cause adopted by Democritus is true; I shall say only that he seems to me not entirely refuted by Aristotle, who says that if it were true that the hot atoms which ascend sustained heavy bodies that are broad, they ought to do that more in air than in water—perhaps because in Aris- [130] totle's opinion the same hot corpuscles ascend through air with greater force and speed than through water. If this is, as I believe, [the ground of] Aristotle's objections, it seems to me we have cause to ask whether he may not have been deceived on more than one count.

First: of those hot things, whether igneous corpuscles, or exhalation, or in a word whatever material these may be that also go up in air, it is not credible that they rise more speedily in air than in water; on the contrary, perhaps they move more impetuously through water than through air, as I have partly demonstrated above. Here I know not how to discover the cause that Aristotle, seeing that downward motion of the same moveable is swifter in air than in water, did not notice here that the opposite must happen for the contrary motion; that is, that it must be faster in water than in air. For inasmuch as the descending moveable must move more swiftly through air than through water, then if we imagine its heaviness gradually diminished, this will at first be such that falling swiftly in air, it will sink very slowly in water, and next it may become such that though it still falls through air, it will rise in water; then, made still less heavy, it will ascend swiftly through water and yet will still fall in air—and, in sum, before it begins to be able to ascend in air, however slowly, it will very swiftly rise up through water. How, then, is it true that what moves upward moves more swiftly through air than through water?

What made Aristotle believe that motion upward is made more swiftly in air than in water was, in the first place, his having related the causes of the slow and the swift, as well for motion up as down, solely to differences of shape of the moveable and to the greater or less corporeality or subtlety of the medium, neglecting comparison of excess of heaviness in the moveables and the mediums, which yet is the most principal point in this business. For if increase or decrease in slowness or speed were related only to the grossness or subtlety of the mediums, every moveable that fell through air would also descend through water, since whatever difference was found between the corporeality of water and that of air could very well be found between the speed of a given moveable in air and some other speed, and this would then have to be its own [speed] in water, which is nevertheless most false.

The other occasion [of Aristotle's belief] was that he believed [131] that, just as there is an inherent and positive quality by which elemental bodies have a propensity to move toward the earth's center, so there exists another, no less intrinsic, by which some bodies had an impetus to flee the center and move upward, by virtue of this intrinsic principle he named *levity*. Hence moveables with that motion would more easily fend thinner than thicker mediums. But that position likewise is shown to be insecure, as I hinted before in part, and as I could show by reasons and experiences if the present occasion had greater need of that, or if I could explain myself in a few words.

Simplicio Here I should have preferred a fuller discussion of this matter, both as to our friend's conviction that Aristotle was guided in this matter by such considerations, and as to the reasons and experiences against it. But I believe I have caught his drift from what has gone before, about those ratios of speeds.

Salviati Without wasting many words, I can mention the experience of an inflated bladder held below water and released. This indeed moves upward in water with great speed, but then on entering the air is swiftly brought to rest and falls back.

Simplicio But if Aristotle was thinking along the line alleged, he clearly would have been alluding to things that move upward in air, not those that fall in it.

Salviati Very well, and our friend would reply that if we imagine the bladder filled not with air, but with air mixed with fire particles until it became even lighter than air, then it would indeed rise in air, though slowly, while it would have risen through water even more speedily than when filled with air alone.

Sagredo I have not had much to say in this session, but what Salviati has just said awakens in me an old question I once asked our friend. In discussing certain experiments we had performed, concerning the force of percussion as I recall it, he argued similarly from imagined experiences in addition to those we had seen. I remarked that in so doing it seemed to me he was violating his own principle of consulting nature directly. He replied that indeed we should do that whenever we can, and should accept no higher authority; but when we cannot, we should extend what we have seen to other, unobservable cases using common sense and reason. The purpose of sensible experiments, he said, was to make sure of our facts, while the purpose of imagined experiences was to make sure of our language in describing them.† The facts would always remain consistent, but some ways of describing them might lead to inconveniences or to contradictions, while other ways would not do so and would reduce our puzzles to a minimum. In science, he said, we should adopt the most consistent ways of talking, though without supposing our words could thus gain any power over facts. For, as he also said, "things come first, and names afterward."

Simplicio Thank you both for these hints; now, to go on:

Therefore Aristotle's objection against Democritus is not good when he says that if the ascending igneous atoms sustain heavy bodies of broad shape, this should happen more in air than in water because such corpuscles move more rapidly in the former than in the latter. Rather, the exact opposite should happen, since they ascend more slowly in air, and not only that but they go not united together as in water, but separately, and they scatter, so to speak, whence, as Democritus well said in replying to and resolving his own objection, they do not go unitedly to strike and make impetus. When Aristotle will have it that the said heavy bodies would be more easily sustained by ascending hot [particles] in air than in water, he is mistaken in the second place by failing to notice that the same heavy bodies are so much heavier in air than in water that some body weighing ten pounds in air will not weigh half an ounce in water. How, then, can it be easier to sustain this body in air than in water?

Conclude, then, that in this particular Democritus philosophized better than Aristotle. But I do not thereby affirm that Democritus philosophized correctly;† rather, I shall say that there is a manifest experience that destroys his reasoning. This is that, if it were true that ascending hot atoms in water would sustain a body that without their opposition would go to the bottom, it would follow that we could find some material, little heavier than water, that, when reduced to a ball or other compact shape, would go to the bottom as something that encountered few igneous atoms and that then, expanded into a wide thin plate, would come to be suspended above by the impulsion of a great multitude of those corpuscles and [132] later to be held at the skin of the water surface. This is not seen to happen, experience showing us that a body of spherical shape, for example, which only with great slowness goes to the bottom at all, remains there and will again sink when reduced into any broader figure.

Simplicio Here, then, our friend asserts that the same experiences that have concerned him throughout his book, and that in his view condemned the position of Aristotle, no less condemn and destroy the doctrine of Democritus. Great philosophical paradoxes are thus born, in which neither of two contradictory views is true and both are false; and, moreover, of two writers who are both wrong and who oppose one another, it can be said that one has reasoned better than the other. It seems to me that with such doctrines, our friend's proposed science threatens anarchy and no less than total destruction of philosophy.

Salviati On the contrary, our friend holds that his kind of science should enable us to philosophize better. The key, Simplicio, lies in his view, borrowed from venerable antiquity, that it is far easier to detect falsehood than to discover truth. Aristotle repeatedly discovered error in the arguments of his predecessors, and even in his great teacher, Plato, showing that even the greatest thinkers had erred. But that Aristotle invariably arrived at truth is not thereby established; on the contrary, it is probable (by the same rule) that he too occasionally erred, leaving some little room for the ages that followed to detect those errors and set them straight.

That is the purpose of our friend's new science as he sees it; and, just as the jeweler's balance is more accurate than the merchant's steelyard, so we may expect in time to have scales as much more delicate than the former as that is than the latter. We may not yet know the exact weight of anything, but we do not therefore doubt that everything has its weight, or that in years to come we shall be able to improve our weighings, without every arriving at the final, exact, and absolute weight of any single thing in the universe.

Simplicio This is an old dogma of the skeptics, injurious to philosophy, that all things are relative.

Salviati Not that *things* are, our friend avers without dogma and without skepticism, but that *our knowledge of them* is, and may ever remain so. That is hardly the same in its implications for philosophy.

Simplicio It is a clever distinction, but I fear it may be one without a difference. The cleverness of skeptics often passes for profundity. Well, our text continues:

> It is therefore necessary to say . . .

Here the word *say,* Sagredo, reminds me that you stressed our friend's concern with various ways of describing facts, quite apart from the essential nature of facts themselves. Here he explains how we must speak, not how things must be. Perhaps this is much like the distinction we were just discussing, and I too hastily rejected it.

> It is therefore necessary to say either that in water there are no such ascending igneous atoms, or that if there are, they are not capable of lifting and pushing up any plate of material that without them would go to the bottom. I deem the second of these two positions true, supposing water in its natural state of coldness. But if we take a vessel of glass or copper or any other hard material, filled with cold water, in which is placed a solid having a flat or concave shape, which exceeds water in specific gravity so little that it goes slowly to the bottom, I say that when under this vessel are

placed burning coals, then as soon as new igneous corpuscles have penetrated the substance of the vessel, they ascend through that of the water; and doubtless in striking the aforesaid solid they will push it clear to the surface, and will hold it there as long as the incursion of the said corpuscles shall endure; but on the cessation of this with removal of the fire, the solid will return to the bottom, abandoned by its little points [of support]. But let Democritus note that this cause exists only when we are dealing with the raising and sustaining of plates of material little heavier than water, or exceedingly thin, while in heavy matter of some thickness, such as plates of lead or other metal, any such effect entirely ceases. In witness of this, note that such plates [as are actually] raised by igneous atoms ascend through the whole depth of the water and stop at the airy boundary, remaining [just] below water, whereas the adversaries' chips do not stop [sinking] except when their upper surfaces remain dry; nor is there any way, once they are within the water, of acting so that they do not fall to the bottom. So the cause of the swimming on water of the things Democritus speaks of is different from that of the things we speak of.

But, going back to Aristotle, it seems to me that he less warmly refutes Democritus than, according to Aristotle's statement, Democritus refuted the objection he had raised against himself. This opposing him by saying that if the ascending hot [particles] were what raised the thin plate, a solid should be even more supported and raised through air, shows in Aristotle a greater desire for the overthrow of Democritus than for exactness in sound philosophizing. [133] That same desire is found on other occasions—and, without looking too far from this place, in the text [immediately] preceding this [last] chapter [of *De caelò*] that we have in our hands. There Aristotle tries to refute this same Democritus, who, not content merely with names, wished to explain more particularly what *gravity* and *levity* were; that is, [to explain] the cause of going down and that of ascending. For this he introduced the plenum and the void, giving to fire void by which it is moved upward, and to earth fullness by which that descends, attributing then to air more of fire, and to water more of earth. But Aristotle, wishing also a positive cause for motion upward—and not, like Plato and these others, a simple negation or privation—argues against Democritus and says: "If what you suppose were true, then there will be a great bulk of water that will have more of fire than does some small bulk of air, and a great mass of air that will have more earth than some small mass of water, whence necessarily a great bulk of air must come down faster than some small quantity of water. But that is not seen in any case; therefore Democritus reasoned erroneously."

But in my opinion the doctrine of Democritus is not overthrown by this objection; rather, if I am not mistaken, Aristotle's mode of deduction is either inconclusive, or if conclusive it can be turned against him. For let Democritus grant to Aristotle that a great mass of air *can* be taken that contains more earth than does some little quantity of water, but deny that such a mass of air must go down faster than a little water, and for many reasons. First, because if the greater quantity of earth contained in the great mass of air could be the cause of greater speed than the lesser quantity of earth contained in the little bulk of water, then in the first place it would have to be true that a greater bulk of simple earth is moved more swiftly than a lesser amount; but this is false, though Aristotle affirms in many places that it is true.

Simplicio I do not know what to make of this. Aristotle does say in many places that the heavier moves downward more quickly than the light, and that is not in any way false, but most true and apparent. A cork, for example, descends less swiftly than a bit of lead, as we can see and test any time.

Salviati Our friend did not deny this when he said it is false that a greater amount of earth moves more swiftly than a small amount.

Simplicio But in the cork there is a small amount of earthy substance, with much air, and in the lead a great amount with little or even no air, which is why the cork moves less swiftly.

Salviati How do you know that there is little earth in cork, and much in lead?

Simplicio In many ways. One way is to subject both to fire, which drives out all air and water, so that the ash remaining in the case of cork is all the earthy matter it contained. But the lead remains unchanged by fire.

Salviati Will not this remaining ash fall even less swiftly than the cork?

Simplicio Indeed it will, and if pulverized it will remain floating in air, as Aristotle said of dust—though I recall that our friend wishes also to deny that.

Salviati Well, if all air and water are driven away by fire, and only earthy matter remains in cork ash, how can that earthy matter fall less swiftly in air than cork itself, which contains this same earth and also water, which falls in air?

Simplicio Very easily, for this earthy matter is very small in total bulk, whereas that of the lead is large; and the small falls faster than the large, just as Aristotle said.

Salviati But this is to prove idem per idem; you assume of earth that a small amount falls less swiftly than a large, and to prove this you offer it as your explanation. Now, our friend puts the matter to the test, not by supposing there is earthy matter in cork but by taking earthy matter, such as a brick, and dropping it alongside half a brick, which has half as much earthy matter. Aristotle says the brick should fall twice as fast as its half, whereas in a test from a high tower it does not, but they reach the ground together.

Simplicio That is quite impossible.

Salviati Let us see. Here are two bricks; with your permission, Sagredo, I shall break one in half. Holding a brick in one hand, and a half-brick in the other, as high as I can, I shall drop them together. You say that the brick should reach the ground when the half-brick is but halfway there. Now I drop them. You see that both strike the ground together, as nearly as we can judge.

Simplicio Dear me. But in so short a distance, both travel so swiftly that we are uncertain that they were truly dropped, and struck the ground, at the same times.

Salviati Our friend has demonstrated this through great heights, and long ago he proved it to be necessary. Before his time the same had been proved by Giovanni Battista Benedetti, a great mathematician, after which it was tested carefully by Simon Stevin through a height of thirty feet. So on this point, Simplicio, Aristotle was simply mistaken, and you are only persuaded differently by arguing circularly, as I remarked before.

Sagredo I too can assure you, Simplicio, that when our friend first told me his conclusion I, like you, thought it impossible; yet, as long as both bodies are of the same material and are not excessively small, the result is always the same. I think we should spend no more time on this matter now, but go on to see how our friend reasons from this sensate experience.

Simplicio First I must hear his "necessary demonstration," for you say his science requires both in all cases.

Salviati He argues thus. Two half-bricks, released simultaneously, will fall through air at equal speeds. Therefore neither one could pull on the other if they were suddenly connected during fall. Consequently a brick, which is merely two half-bricks connected, will fall no faster than either half alone.

Simplicio The argument is clever, though I suspect some fallacy. I need time to think about, it, so now I agree to proceed.

It is not greater absolute heaviness, but greater specific heaviness, that is the cause of greater speed, nor does a ball of wood weighing ten pounds descend more swiftly than one of the same material that weighs ten ounces.

Simplicio There, you see; already the fallacy is apparent. You remember how our friend adduced the settling out of earthy particles from muddy water as evidence that water has no resistance to division. The descent of all such particles, he said, might take four or six days. But a lump of earth will go to the bottom of the same water in less than one second of time, and its specific heaviness is the same as that of any earthy particle, however small. I do not see how it can be anything other than greater absolute heaviness that is the cause of greater speed in this case, reasoning from examples supplied by our friend himself. This time I may indeed say that his science is "far from exact," Sagredo, though I should not have said that before on the matter of compression of air.

Sagredo Once again, Simplicio, your haste to contradict our friend's position has not left him time to complete its exposition. You say you do not see, now, how difference of downward natural speed can be accounted for by anything but either absolute heaviness or specific heaviness, those being the only two things yet mentioned as possible causes, and our friend having eliminated one of them. But there is something else to be considered, something that is a matter of simple geometry rather than of heaviness of either kind, not considered by either Democritus or Aristotle, and still less by his supporter Buonamico. Our friend will come to this in due course, after clearing away various misapprehensions to bring this matter of natural downward motion into logical perspective. So for the present I ask you to hold in your mind a reservation, remembering that there is a difficulty that you have well and clearly expressed, so that our friend is obliged before he is through to meet it head on or else to admit defeat in the present argument. It is not possible, in presenting a new science that will replace old principles, to deal with everything at one fell swoop; things must be taken up one at a time in orderly sequence. As Aristotle well said, and you yourself, Simplicio, cited earlier in our talks, it is wrong to attack old principles unless one has better ones to put in their place. By the same token, it is wrong to refuse to listen to all the new ones and insist that each in turn be rejected because it, by itself, cannot be reconciled to old principles.

Simplicio I understand what you are saying, though as yet I do not see how our friend's position thus far can be rescued from patent contradiction, and least of all how it can be rescued by mere geometry, which has no proper place in physics, geometry being abstracted from all material—and grains of dust or chunks of earth are nothing if not material. Still, as a philosopher, and certain of the rightness of my judgment, I can afford to allow our friend all the rope he needs to hang himself. I shall accordingly do as you say; and, allowing his argument from the two half-bricks to establish the equality of their speed of fall with that of a whole brick, I shall go on reading until we come to some geometrical reason why a grain of brick dust does not descend beside a falling brick. But already I see that this must have something to do with another cause, outside the material of bricks, which cannot be the specific heaviness of that material and *may* not be simple absolute heaviness, though it must be somehow related to it. Our friend proceeds thus, in this place:

But it is true that a four-ounce lead ball descends more swiftly than a twenty-pound wooden ball, because lead is specifically heavier than wood. Hence it is not necessary that a great mass of air, by the [134] much earth contained in it, should descend more swiftly than a small bulk of water; on the contrary, any bulk of water must move more swiftly than any other of air, there being a greater share of the terraneous part, specifically greater [in heaviness], in water than in air.

Note in the second place that, in multiplying the bulk of the air, there is multiplied not only what there is of earthy [substance], but also its fiery part as well, whereby the cause of going upward, by virtue of fire, is no less increased than that of going down on account of the multiplied earth. In increasing the magnitude of the air it would be necessary to multiply only what it has of earth, leaving its fire in the original state; for then, the augmented earthiness of the air surpassing the earthy part of the small quantity of water, it oould have been possible to claim with more probability that the great quantity of air ought to descend with greater impetus than the little water.

Therefore the fallacy is rather in Aristotle's reasoning than in that of Democritus, who could impugn Aristotle with a like reason, saying: "If it is true that the extreme elements be one simply heavy and the other simply light, while the middle ones share in both natures, but air more in lightness and water more in heaviness, then there will be a great bulk of air whose heaviness exceeds the heaviness of a small quantity of water; whence that mass of air will *descend* more swiftly than this little water; but that is never seen to happen; therefore it is not true that the middle elements share in both qualities." Such an argument is fallacious no less than that other one [in Aristotle] against Democritus.

Simplicio I must confess that when I first read this passage last night I was for a time bewildered, for it is true that formally the argument our friend offered to Democritus as a reply to Aristotle is precisely the same as the argument Aristotle made against Democritus on this point; yet I thought there must be some trick, for there had seemed to me no fallacy in Aristotle's reasoning. And, if there is a trick, it might be just the usual trick of Sextus Empiricus and the other skeptics by which they pretend to show that there is never a better reason for holding one view than for holding its opposite. Yet we know that our friend is no Pyhrronist; indeed, his main fault is too great a certainty that he is in the right. The truth seems to be that both Aristotle and Democritus neglected that relative weight of moveable and medium, first brought in by Archimedes. The arguments on both sides of this dispute therefore remained inconclusive, by reason of an omission common to both positions. On the relevance of that relative heaviness of moveable and medium, our friend rests his whole case. I do not say he is right, any more than he says Democritus is right in supplying him with a new reply to Aristotle; indeed, he concludes by remarking that the argument is fallacious. But I do grant that whether or not our friend is right about the cause of floating, he has exposed a new consideration in philosophical disputations, one of special value in replying to skeptics. This is that when it appears that as good reason exists for one view as for its opposite, we may suspect some oversight on both sides, something left out that has to be taken into account. As an Aristotelian, I question his search for this merely in sensible experience, which can never guarantee that we have taken everything relevant into account, and this may be only a new kind of skepticism more pernicious than the old. Until something more profound than a doctrine of solids in water emerges from it, I place my faith in Aristotle against all his opponents.

Salviati A certain *kind* of skepticism, or rather of suspended judgment, is indeed part of our friend's new science, as is seen from his saying that no event in nature, however small, will ever be completely understood by theorists. But that this new skepticism is pernicious I do not believe, for suspension of final judgment is very different from the fatuous overconfidence of true skeptics. It is one thing to say "my science is true" and quite another to say "my science is true as far as it goes, but it is still incomplete and may never be completed."

Sagredo I recall that our friend, when still at Padua, used to say when students asked some questions: "That is one of the thousand things I do not know," which other professors took as degrading the university and the teaching profession, in which one's duty is to know and answer every question, if only by consulting Zimarra's concordance to the works of Aristotle.

Simplicio I recall these things only too well, and I must tell you frankly that they tended to undermine proper respect and the order of society. Though I was sorry to see our friend leave, things at the university have gone more smoothly since he did. From this book it is apparent that he has carried to the Tuscan court the same spirit of dissension that prevailed around him at Padua. It is one thing for me, as a philosopher, to ruminate on the nature of his arguments, and quite another thing for him to put them into the hands of common people by writing them in Italian. If I am any judge, our friend will eventually come to grief by this tactic. But time flies, and we must get back to our examination of his book, which now proceeds:

Aristotle, having said that if the position of Democritus were true it would follow necessarily that a great bulk of air must move [down] more quickly than a small bulk of water, ultimately added that this is never seen in any way. It seems to me that someone might remain with a desire to hear from him in what place that should happen which he deduces against Democritus, and what experience would teach us that it does not happen. It is vain to think of seeing this [falling of air and water] in the element of water or in that of air, since water neither moves (or will ever move) in water, nor air in air, whatever shares of earth and fire are assigned to them. A still more inappropriate place for any such experience is earth, that being not a fluid body that yields to the mobility of other bodies. By Aristotle's own dictum, the void does not exist, and if it existed nothing would move in it. There remains only the region of [135] fire, but since that is quite a distance away from us, what experience will be able to assure us, or can have assured Aristotle in the way he needs—as a thing well known to the senses—in affirming what he produces in refutation of Democritus; that is, that a great bulk of air does not move more swiftly than a small bulk of water?

But I do not wish to remain longer on this subject, concerning which much more could be said; and, putting aside Democritus, I return to the text of Aristotle. Here he girds his loins to render the true causes whence it comes about that thin plates of iron or lead swim above on water, and that even gold itself, thinned into very tenuous leaves, and fine powder, swim not only in water but even in air. He assumes that some continua are easily divisible and others not, and that among those easily divisible some are more and others less so; and these things, he declares, must be esteemed to be the causes sought. He then adds that things easily shaped are easily divisible, and the more easily so the more readily shaped, whence air [changes shape and is divided] more [easily] than water, and water than earth. Finally, he supposes that in every class, a lesser quantity is more easily divided and separated than a greater.

Here I note that Aristotle's conclusions in general are true, but it seems to me that he applies them to particulars in which they have no place, though they do in others. For example, wax is more easily divisible than lead, and lead than silver, just as wax more readily receives any boundary than lead, and lead than silver. It is moreover true that a small quantity of silver is divided more easily than a great mass; and all these propositions are true because it is true that in silver and lead and wax there is simple resistance to being divided, and where there is the absolute there is also the relative. But if in water, as in air, there is no resistance at all to simple division, how can we say that water is divided with more difficulty than air? We cannot get free from equivocation thus, so I again repeat that it is one thing to resist absolute division, and another to resist division with such-and-such a speed. But in order to produce rest and to stop motion, resistance to division must be absolute, while resistance to swift division is the cause not of rest, but of slowness of motion. Now it is manifest that in air, as in water, there is no resistance to simple division, for no solid body is found [136] that does not divide both air and water.

To say that gold leaf or fine powder lacks power to overcome resistance of the air is contrary to what experience shows us, gold and dust being seen wandering [down] through air and finally settling to the bottom, doing the same in water if they are situated there and separated from air. Now since, as I say, neither water nor air resists simple division at all, it cannot be said that water resists more than air. Nor let him who opposes me give the example of very light bodies, like a feather or a bit of pith taken from cornstalks or marsh grass, that fends air but not water, and infer from this that air is more easily divisible than water. For I shall reply that if one looks closely he will see the same solid also divide the continuity of water and partly submerge—by a part such that an equal volume of water would weigh the same as it.

And if he still persists in arguing that that solid goes [not] down through impotence to divide water, I shall tell him to push it under water and then watch it, set at liberty, dividing water in ascending no less readily than it divided air in descending. So that to say, "This solid descends in air, but on reaching water ceases to move; hence water is divided with more difficulty" concludes nothing, and I on the contrary shall propose a stick or piece of wax that rises from the bottom of water and easily divides its resistance, which [object] then, having arrived at the air, stops and hardly touches it. So I, with equal reason, may say that water is more easily divided than air.

Simplicio Here again is that kind of trick that turns an argument against its user, applied by Sextus Empiricus to every position that has ever been put forth, but, as I found and told you earlier, valid only in certain cases. It now strikes me that where it is applicable, and powerful, as here, there seems always to be a kind of symmetry in the assumption on which the original position has been based. In this case the resistance is supposed the same to motion of any kind, downward or upward, and the motions named are also symmetrical and contrary, so that nothing can be said of one without equal implications for the other. The original argument has then simply overlooked those implications, its author being intent on establishing a particular motion; and the trick is merely to fill in what he neglected. So, as our friend observed, either the original argument is inconclusive, or it is conclusive equally on both sides of the debated point. Perhaps Aristotle omitted this from his listing of logical fallacies because there is no *logical* fallacy in the original argument, but only an omission or neglect.

Salviati This situation arises, I believe, from certain ways of talking about nature, and it was probably recognized by our friend from his interest in that matter. For just as the ancients wished heaviness alone in all things, and Aristotle changed that way of talking by introducing positive levity, so we are in the habit, following Aristotle, of speaking of coldness as positive, and tardity as positive, and so on, where we could speak only of degrees of heat, degrees of speed, and so on. This habit of speech may tempt us to argue in one direction without noticing either the symmetry of which you speak or the implications that render our conclusions inconclusive—or if conclusive, equally so against us.

Simplicio I must have time to think about this. Meanwhile, the text goes on:

In this matter I do not wish to stop short of pointing out another fallacy of those who still attribute the cause of going or not going to the bottom to less or greater resistance of the corporeality of water against being divided. They introduce the example of an egg, whch goes to the bottom in fresh water but floats in salt water, adducing as the cause of this the little resistance to being divided of fresh water, and the greater resistance of salt water. But, if I am not mistaken, from the same experience can be deduced also the opposite; that is, that sweet water is thicker and salt water more tenuous and thinner; because an egg [also] rises readily from the bottom in salt water and divides its resistance, which it cannot do in fresh water, at the bottom of which it rests without being able to lift itself up. To like straits will false principles lead; but whoever philoso- [137] phizes correctly, and recognizes as the cause of such effects excesses of heaviness of moveables and of mediums, will say that the egg goes to the bottom in fresh water because it is heavier than that; and without any obstacle he will very solidly establish his conclusions.

Salviati Allow me to interrupt here, Simplicio, to elaborate on the point we were just discussing. Note that our friend here says *less heavy,* and not *lighter.* Now, *less heavy* and *more heavy* are not less symmetrical and contrary than *heavier* and *lighter,* but the form of speech chosen tends to reduce the reader's temptation to see lightness, or levity, as a positive quality and inclines him to think instead in terms of degrees of heaviness alone. In these cases, where we must compare not just the heaviness in two different bodies, but also the heaviness in either with the heaviness of the medium, the manner of talking may assist greatly in avoiding confusion, by which some symmetries with unexplored implications might remain. Indeed, it may be that Aristotle himself was prevented from noticing his omission of the important principle introduced later by Archimedes only through Aristotle's habitual treatment of both levity and gravity as simple positive qualities. That manner of speaking, though blameless in itself, makes it more difficult to distinguish the absolute from the relative cases, as is here necessary for clarity.

Simplicio Until today I should have stoutly resisted your notion that Aristotle left anything unclear, let alone suffered from any confusion of qualities. Now, having pondered this very matter last night, reading over this section, I realize now that *if* Aristotle did omit anything, then yours is the best way of explaining that. But to resume:

Therefore the reason that Aristotle adds in the text must go, when he says, "Things, therefore, that have great breadth, remain on top because they include much, and what is greater is less easily divided." Let such reasoning cease, I say, because it is not true that in water or in air there exists any resistance to division; besides which, the leaden flake, when it does stop [descending], has already divided and penetrated the corporeality of the water and has entered into it ten or twelve times as far as its own thickness. Furthermore, if there were any such resistance in water to its being divided, it would be simpleminded to say that this existed more in the upper parts than in the middle and lower parts; rather, if there had to be a difference, greater corporeality should exist in the lower parts, so that the flake should be no less unable to penetrate the lower than the upper parts of water. Yet we see that no sooner is the upper surface of the lamina wetted than it precipitously and without any stay descends clear to the bottom.

I do not believe that anyone would say (thinking perhaps in such a way to defend Aristotle) that, it being true that much water resists more than little, the said lamina, placed lower, would descend [merely] because a less bulk of water remained to be divided; for if, after having seen the same flake float above four inches of water and later become submerged in it, he tried the same experience over a depth of ten or twenty feet, he would see precisely the same effect. And here, to remove a very common error, I shall again mention that a ship, or any other body, that floats above a depth of one hundred or one thousand feet, dipping in only six feet of its own height, will float exactly the same way in water no deeper than six feet one-half inch. Likewise, I do not believe it can be said that the upper parts of water are crasser, though a serious author has deemed the topmost water in the sea to be such, taking his argument from its being found more salty then deeper water. In this experience I question whether in extracting water from the bottom [138] there was not encountered some spring of fresh water that spouted there. Also we see, on the other hand, that the fresh water of rivers spreads out for some miles beyond their mouths upon the salt water of the sea, without descending or mixing with it unless there is some commotion or disturbance by winds.

Now, getting back to Aristotle, I say that breadth of shape has not even a little to do with this business; for the same plate of lead or other material, cut into strips as narrow as you like, swims neither more nor less on top; and the same holds when the said strips are cut again into little squares, because not breadth, but thickness, is what operates in this matter. I tell him, moreover, that if it were indeed true that opposition to division were the proper cause of floating, narrow and short shapes would float much better than those larger and broader. Hence, by increasing the breadth of the shape one would diminish the ease of its floating, and decreasing the former, one would increase the latter [ad absurdum].

Sagredo I interrupt here, Simplicio, to remind you of our debate earlier, when you said that absolute heaviness must be the reason that particles of mud settle out from turbid water very slowly, whereas chunks of earth, though of the same specific heaviness, go to the bottom swiftly. Our friend has now arrived at the point at which, having cleared away many plausible controversial points, he can bring in pure geometry to explain how speed of natural downward motion through a medium is *affected* by absolute heaviness, though one cannot say that its *cause* is absolute heaviness or that the speeds are proportional to the weights, as Aristotle appears to have assumed or believed. For the effect is not that of bulk, or weight, but of the ratio of bulk or weight to something else with which it is necessarily connected by simple geometry; and ratios do not follow absolute quantities, but follow the relations existing between them. I am sure you now see, or will see as you proceed, why our friend did not attempt to deal with this consideration at the very beginning, as if his science were nothing more than pure mathematics, but first explained the physical aspects of solids placed in air or water and allowed to move freely there, saving the geometrical reasoning for the end and final clarification of his approach to this ancient problem of shapes and speeds in descent.

Simplicio I grant that if he had begun with pure geometry, I should have been even less able and willing to listen to his arguments than I am now, for by now I am very curious to hear his final explanation, whereas earlier I should merely have been put off by that show of mathematical knowledge that seems to Platonists sufficient to silence all opposition. Here, then, is what our friend has to say:

And to explain what I say, I shall put it into consideration that when a thin plate of lead descends, dividing the water, the division and separation is made between parts of water that are very near the perimeter and circumference of this plate, and, according to the greater or less size of that circuit, greater or less quantities of water have to be divided. Thus if the circuit of a board shall be, for example, ten feet, then in going down flat it must make a separation and division, and, so to speak, a cut, of ten feet in length in water. Similarly, a smaller plate that has a perimeter of four feet must make a cut of four feet. That being the case, anyone who knows a bit of geometry will understand not only that a board cut into many strips will float much better than when it was whole, but that all shapes must better float as they are shorter and narrower. Let the board ABDC, for example, be eight inches long and five broad; its circuit will be 26 inches, and that will be the length of the cut that it must make in water to descend through it. But if we slice it into eight smaller boards along the lines EF, GH, and so on, making seven cuts, we shall be adding to the 26 inches of circuit for the whole board

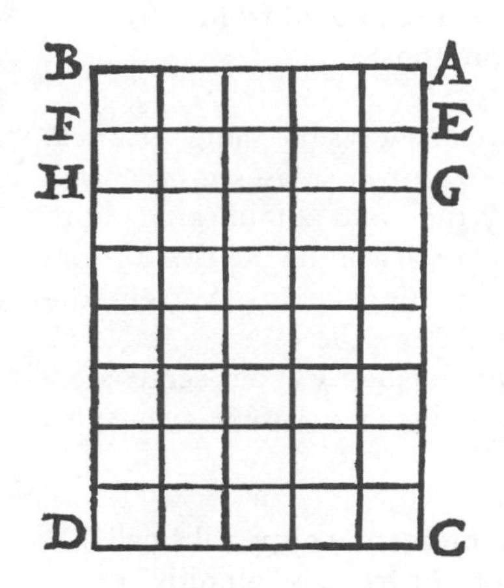

[139] another 70, whence the eight small boards sawed thus and separated, will have to cut 96 inches of water. If now we saw each of the small boards into five parts, reducing them to squares, we shall add to the 96 inches of circuit, with four sawings of eight inches each, another 64 inches, whence the said squares, to descend in water, must divide 160 inches of water. The resistance of 160 is much greater than that of 26; therefore the smaller we make the surfaces, the easier it is for

them to float. The same happens for all other shapes whose surfaces are [geometrically] similar but that differ in size; for diminish or increase the said surfaces in size as you will, their perimeters always diminish or grow as the square roots, and with them the resistances they meet in fending water. Therefore plates and boards float more easily bit by bit, according as they are of smaller size.

Salviati It should be noted that this statement, though it appears absolute, is governed by the beginning of this mathematical derivation, where our friend said "if it were indeed true that opposition to division were the proper cause of floating . . . ," for the entire derivation assumed "cuts" to be made in water.

Simplicio Yes, I know, and though I was in some doubt of this at the place just read, it was cleared away by the final sentence below, which in the first edition was at this place but was separated from it by a paragraph added to the second edition, thus:

This is manifest because if we maintain always the same thickness [*altezza*] of the solid, the base increases or decreases in the same ratio as the solid [volume]. Hence, the solid diminishing more [rapidly] than the circuit, the cause of its going to the bottom diminishes more than the cause of its floating; and, conversely, the more the solid increases in circuit, the more the cause of its going to the bottom increases and that of its remaining afloat decreases.

And all this would follow in the teaching of Aristotle, against his very own doctrine.

Sagredo Salviati, since you agreed with me in preferring a special small resistance at the surface of water over our friend's "affinitiy" or "magnetic adhesion" of air to dry surfaces, we can make a use of his method of geometric ratios here that he could not, applying it to the floating of dust and small filings on water, whereas for him it served only to explain the very slow settling of fine particles.

Salviati So we can. Thank you for pointing out this interesting fact, which had escaped me. It illuminates the concluding sentence above, to which Simplicio called attention, of which the idea was that "all this would *still* follow under Aristotle's teaching," which was contradicted by showing that very small solids would more readily float, and not more readily pierce the supposed resistance of water to division. By the same token, it will still follow if at some future time it is discovered that a measurable, small resistance to penetration does exist at the surface, and only at the surface, of a fluid. In that event this mathematical investigation by our friend would remain of value, and would even take on new value, if (as you and I are inclined to believe) it should become necessary to cast out from science his explanation of the adhesion of air to dry surfaces as holding back the little ridges of water.

Simplicio Naturally I do not accept this mathematical gambit as a valid refutation of Aristotle's physical conclusion that fluids, like solids, resist division, but less strongly than solids. Yet I see that you two, who admire our friend's science perhaps as much as I admire Aristotle's, do not hesitate to anticipate the future downfall of parts of it. This increases the distress I mentioned once before at the prospect that a secure and unshakable natural philosophy, such as we now possess, may be followed by a shifting and uncertain so-called science founded only on the senses and mathematics, and exposed to criticism by mere artisans and reckoners. Well, our friend now concludes his book as follows:

Finally, we read in the last part of the text [of *De Caelo*] that heaviness of the moveable must be compared with resistance of the medium against division, because if the power of the former exceeds the latter it will sink, and, if not, it will swim above. Nothing more need be said on this than has been said already; namely, that it is not resistance to simple division, which does not exist in water or air, but heaviness of the medium that must be compared with the heaviness of the moveable. That being greater in the medium, the moveable will not descend in it, nor even submerge entirely, since in the place it occupies in the water there cannot rest a body weighing less than as much water; but if the moveable shall be heavier, it will descend to the bottom, to occupy a place where it is [140] more suitable to nature for it to rest than some less heavy body.

And that is the single, true, proper, and absolute cause of swimming above or going to the bottom, so that no other [cause] plays a part in this.† And the adversaries' [ebony] chip will swim when it is coupled with as much air as, together with it, will form a body less heavy than as much water as would go to fill the place occupied in the water by this composite. But when simple ebony shall be placed in water, in conformity with the tenor of our question, it will always go to the bottom, though it be as thin as paper. [End of text.]

Sagredo Well, we have come to the end of our watery voyage, alternately floating on clouds of airy speculation and sinking to the bottom to explore their foundations, sometimes becoming a bit giddy in the former and nearly drowning while we were about the latter. As usual I have gained new knowledge by rereading our friend's arguments, and even more from the information that you, Salviati, have contributed about the circumstances of the debate, as also from the acute criticisms and objections that you, Simplicio, have raised as a philosopher. I am obliged to you both.

Salviati No less am I obliged to Simplicio for many reflections that would otherwise have escaped my consideration, as well as for some that I know our friend will be glad to hear when I write to him; and to Sagredo I am indebted for his splendid hospitality that makes me reluctant indeed to embark on my new, more literal, voyage.

Simplicio And my obligation to both of you is redoubled by your having taken me away from the tedious work of polishing my book for publication. Though I lack your skill in following mathematical demonstrations and your familiarity with actual experiments, I have not participated in these discussions without benefit. For one thing, I have come to understand how the excitement of discovery may draw men like our friend away from the more seemly pleasures of contemplative thought. For another, I have gained a form of argument that may be of use to me in upsetting skeptics, of whom few are fortunately left in these days, as well as in refuting the sciolism of unsound interpreters of Aristotle, of whom there are only too many. But above all I relished this taste of the old days when our friend was here in person to stir new thoughts, as you have so well done here on his behalf, Salviati.

Salviati Thank you, Simplicio. When the dispute at Florence was at its height, he used to tell me privately that the quality of his opponents left much to be desired in comparison with those he had had at Padua. Your name especially was mentioned, though he warned me that I should find you no less opposed to his new science of physical investigation than you were to his innovations in astronomy.

Simplicio It is true that I remain as unconvinced that this should be called a science as that those new things he wants to place in the heavens are really there. I believe that true science cannot be gained piecemeal, by his kind of "demonstrative advance," but must proceed from indubitable principles, established by Aristotle in his *Firt Philosophy,* or metaphysics.

Sagredo I, on the other hand, am persuaded that it is precisely from supposing that any principles are truly indubitable, execpting only those of our Catholic faith, that new discovery has been impeded in the past. I think of that doctrine that all heavenly motions must be perfectly circular, with the earth at the exact center, by which Eudoxus accounted for the appearances. So long as that doctrine remained indubitable, we had no reliable tables of such motions; but when Hipparchus and Ptolemy allowed eccentric circles and epicycles, astronomy begain to flourish and became exact and useful, not just schematic and abstract. Might it not be that physics, by departing only slightly from schematic principles, could in time become a useful and not just a speculative science?

Simplicio I do not see how; and even if it could, utility would do no more than enhance *techne,* without increasing *episteme.* Since it is the latter alone that we should call "science," you may see from this why I earlier withheld that name from our friend's enterprise.

Salviati I should say that without applying ourselves to *techne,* we cannot know whether or not it can advance *episteme.*

Sagredo At any rate, Simplicio, it costs philosophers nothing if other people wish to engage in these attempts, even though you may be convinced they cannot aid science. Why, then, do philosophers oppose them?

Simplicio We do not, except when and to the extent that innovations contradict Aristotle; it is then our business to defend him.

Salviati I should have thought, from the very meaning of "philosophy," that your business was to promote knowledge.

Simplicio It is, and since no true knowledge can contradict Aristotle, you see why it is that we intervene.

Sagredo Even in such trifling and unphilosophical matters as floating?

Simplicio Small errors in the beginning lead to great ones at the end.

Salviati Our friend thinks Aristotle made certain small errors at the beginning of natural philosophy, or at least believes it possible that he did. If so, we must be free to inquire and to publish our findings.

Simplicio So you should, as we should be free to expose your errors; no true philosopher will ever contend otherwise.

Salviati I fear you are mistaken, Simplicio, for even now a band of philosophers at Pisa and Florence† has vowed not just to defend Aristotle, but to oppose everything our friend may say.

Simplicio I much regret to hear this, and shall say first that such men are not entitled to be called by the honorable name of philosopher, and second, that to oppose is not to impede freedom of inquiry, or even of publication.

Salviati In the present matter I have reason to believe it a first step in that direction, those who have published replies to this book being now engaged in threatening any printer who would publish his rejoinders.†

Sagredo This surprises me less than you, Simplicio, for I warned our friend that he would encounter at Florence opposition of a kind he never met here in the freer air of Venice. But now it is time to end our colloquies and repay our guest with more sights of our city before he sails for Spain. Do not forget, Salviati, that we hope upon your return to meet again and discuss that other book, on sunspots.

Simplicio And when you write to our friend, convey to him my most hearty greetings.

Sagredo And my best wishes for whatever new investigations he undertakes. Well, gentlemen, our gondola is waiting; let us go.

Appendix

The first section of this appendix sketches the history of hydrostatics to Galileo's time as it relates to the dispute that gave rise to his *Discourse,* and the bearing of that on the later Copernican episode. The remaining sections deal with particular points in each of the four "days" into which my dialogue is divided. By and large, Salviati and Sagredo are supposed to speak from standpoints similar to those of Galileo and his correspondents, judged from letters, notes relating to the controversy, and other writings of Galileo bearing on similar scientific and philosophical issues. Into their speeches, however, I have inserted various bits of information concerning Galileo that might not have been known to them, and conclusions of my own for which there is no specific documentary evidence. Simplicio is supposed to speak in part for Galileo's Florentine and Pisan adversaries as well as for Cremonini at Padua, whose personal views on the hydrostatic controversy must largely be conjectured. In support of my fictional dialogue, some explanations and some citations are offered not in the conventional form of notes, but as commentaries on selected passages quoted (or paraphrased) in boldface type at the head of each comment, in order of their appearance.

References to original documents as published in the Edizione Nazionale of the *Opere di Galileo Galilei,* edited in twenty volumes by Antonio Favaro, appear simply as a roman numeral indicating the volume and arabic numerals indicating pages. When English translations exist, abbreviations of titles of books containing them are used, as follows:

Dial. Galileo, *Dialogue Concerning the Two Chief World Systems,* trans. S. Drake (Berkeley: University of California Press, 1953, 1962, 1967).

D&O S. Drake, *Discoveries and Opinions of Galileo* (New York: Doubleday, 1957).

GAP S. Drake, *Galileo Against the Philosophers* (Los Angeles: Zeitlin and Ver Bruggle, 1976).

GAW S. Drake, *Galileo At Work* (Chicago: University of Chicago Press, 1978).

GS S. Drake, *Galileo Studies* (Ann Arbor: University of Michigan Press, 1970).

Mech. S. Drake and I. E. Drabkin, *Mechanics in Sixteenth-Century Italy* (Madison: University of Wisconsin Press, 1969).

M&M I. E. Drabkin and S. Drake, *Galileo On Motion and On Mechanics* (Madison: University of Wisconsin Press, 1960).

TNS Galileo, *Two New Sciences,* trans. S. Drake (Madison: University of Wisconsin Press, 1975).

HISTORICAL BACKGROUND

The foundations for hydrostatics were laid by Archimedes in a work called *On Bodies Placed in Water.* It was entirely mathematical except for the physical postulate that any small volume situated within a fluid will move unless the pressures exerted on it from every side are equal. The postulate could not be tested by experiment, but Archimedes was able to prove from it that a solid placed in water loses weight in an amount equal to the weight of water it displaces. That could be tested, and it came to be known among later writers on hydrostatics as "the principle of Archimedes." During the Latin Middle Ages an anonymous summary of elementary results obtained by Archimedes was ascribed to him, and in this was introduced the concept of specific gravity as a distinctive weight of each kind of material. It was of particular use to goldsmiths, jewelers, and traders in precious metals and gems.

Niccolò Tartaglia (1500–1557) published in 1543 a medieval Latin translation of the first book (of two) of the Archimedean treatise, and in 1551 he published an Italian translation with a dialogue commentary, accompanied by his own work on the raising of sunken vessels, the diving bell, and other practical applications. The entire Archimedean treatise was first published in Latin by Federico Commandino in 1565, and in the same year the medieval summary was printed, from papers left by Tartaglia, together with his experimental determinations of the specific weights of various materials. Meanwhile, in 1553, one of Tartaglia's pupils, G. B. Benedetti (1530–90), had used the principle of Archimedes and the assumption that buoyancy of air (alone) retards the fall of heavy bodies to prove that bodies of the same material, falling from the same height, should reach the ground together regardless of their weights. In 1585 Benedetti added a brief treatment of the hydrostatic paradox in a form suggesting the hydraulic lift.

Bendetti's early work on hydrostatics was plagiarized by a Belgian whose book came into the hands of Simon Stevin (1548–1620), who with his friend Jean de Groot carefully checked the equal speed of fall experimentally, explained observed experimental discrepancies, and published its confirmation in 1586. In the same book Stevin gave, in Dutch, a thorough theoretical and practical treatment of hydrostatics in which he discussed fluid pressures and a second form of the hydrostatic paradox. Stevin's work, which may be considered the beginning of modern hydrostatics, was translated into Latin in 1605.

Marino Ghetaldi, who had met Galileo during a visit to Padua, published at Rome in 1603 a detailed experimental investigation of the specific gravities of many solids and liquids. Galileo's notes contain a record of some similar investigations probably made about the time of his first scientific essay. Written in 1586, this concerned the construction of the hydrostatic balance and its use in determining the composition of alloys, prefaced by his reconstruction of the procedure of Archimedes in solving a famous early alloy problem. This was soon followed by Galileo's first essays on motion (1587–92), in which, like Benedetti before him, he used the principle of Archimedes as a base from which to discredit the accepted Aristotelian rules for speeds of falling bodies. Galileo is reported to have demonstrated the incorrectness of those rules, as Stevin had done, by dropping two bodies of the same material but different weights from the Leaning Tower of Pisa in the presence of students and their professors of philosophy. It was thus from the classical science of hydrostatics that Galileo drew inspiration for the first of his many attacks against doctrines of motion basic to Aristotle's natural philosophy.

Aristotle himself had written very little concerning the floating or sinking of bodies in water. As in all cases of natural motion (that is, motion spontaneously undertaken by bodies released from constraint), Aristotle assumed that motion through water was resisted by the medium and that floating occurred because of resistance to division on the part of water. Sharp conflict between university natural philosophy and Galilean experimental science came to the fore precisely over the issue of fluid resistance to penetration; resistance exists, but only at the surface, and it is not a resistance to division, as when we apply a knife to butter. Insistence by Aristotelians on such resistance in accounting for the floating of flakes denser than water conflicted with experimental facts developed and analyzed by Galileo.

In 1628 Galileo's former pupil Benedetto Castelli founded the science of flowing water, after which hydrostatics and hydraulics combined and led on to hydrodynamics, by which we understand both motions of fluids and motions of solids in fluid mediums. Doubtless other early contributions have escaped my attention in the foregoing historical sketch, but it is clear that discussion of motion along with hydrostatic science aroused opposition from professors of philosophy as a potential challenge to Aristotle. Appeal to experiment as the means of deciding the issues was approved by them until the results began to go against their interpretation of Aristotle's brief comments on floating—an interpretation, incidentally, that did Aristotle an injustice, as Galileo pointed out. Determined opposition by Galileo led to their banding together to reject anything he proposed, on this or any other subject, and to seek support from theologians. Thus, before Galileo became involved in the battle over Copernicanism, and long before he published the sciences of motion and strength of materials for which he is more deservedly remembered by scientists, Galileo's views in the seemingly unimportant science of behavior of solids in water had turned philosophers against serious experimental science.

Large issues in the cultural history of modern Europe were raised by the suppression of Galileo's celebrated *Dialogue,* his condemnation to life arrest for "vehement suspicion of heresy," and the polarization of religion and science that ensued in Protestant as well as Catholic countries. For that reason it is important to understand the situation a few years before those events developed.

Catholic authorities had not been hostile, or even indifferent, to science as it developed in the universities beginning with the revival of learning in the twelfth century. On the contrary, science was encouraged as an Aristotelian project of finding the causes of natural events. But physics was regarded as a mere branch of philosophy—natural philosophy—and consequently it fell under the jurisdiction of professors of philosophy. Where their opinions encroached on theological doctrine, the church exercised its control, but it seldom directly regulated natural philosophy. Some apparent exceptions in the Middle Ages applied not to scientific issues as such, but to related Aristotelian implications, such as eternity of the world as against Creation and Judgment Day, or the unreality of atoms and the void. On the former the church opposed Aristotle, while on the latter it supported him. Beyond such issues, more metaphysical than scientific, the church had traditionally avoided taking any official scientific position. It had no need to do so, and early church fathers had cautioned against turning any scientific doctrine into an article of faith, lest heretics better informed in worldly knowledge use any errors in such matters to throw doubt on the wisdom of the church in spiritual affairs, which were properly its special province.

Thus, at the start of Galileo's career, investigation of causes in physical science was under the supervision of professors of philosophy, and it is no wonder they were the first to oppose him publicly. The demonstration from the Leaning Tower is a celebrated example, and the main reason why Galileo's appointment as professor of mathematics at Pisa was allowed to expire in 1592 was the antagonism of professors of philosophy. At Padua, in 1604–5, Galileo's dispute with Cremonini over a new star appears for a time to have threatened his reappointment. His announcement in 1610 of telescopic discoveries aroused new philosophical opposition, particularly over the mountainous surface of the moon and the existence of new "planets." Theologians did not enter into those disputes, except indirectly as philosopical objections were pressed by Jesuits, who had perhaps some theological motivation for supporting the Aristotelian dogma of perfect sphericity and unchangeable heavens.

When the dispute over floating bodies began in the summer of 1611, the Copernican issue was still dormant. There is no reason to think that the first vigorous opposition of philosophers to Galileo's physics had anything to do with his ideas in astronomy. It was aroused solely by his contradiction of Aristotle, and the reasons why it so excited professors of philosophy require some explanation.

The question of floating came up not in a scientific discussion, but as incidental in a philosophical debate on questions of condensation and rarefaction, a very serious matter to natural philosophers. Floating had been mentioned by Aristotle only in his refutation of the doctrines of the Greek atomists and was not regarded as of any great importance for its own sake. But Galileo associated ice with the rarefaction of water, reducing its specific weight, whereas Aristotelians held ice to be condensed water, compacted by cold, bringing in one of the primary qualities used in causal explanations for all kinds of physical changes. It appeared to the Aristotelian professors that all good science was threatened by Galileo's arguments. In their anxiety to save it they asserted, mistakenly supposing Aristotle to have believed this, that shape is a governing factor in the floating of ice in particular and of flat bodies in general. Galileo had no difficulty in countering their arguments and adducing various experiences that distinctly favored his own Archimedean account.

At this stage it would have been possible, and not unprecedented, for the philosophers to reexamine the text of Aristotle, discover that it did not compel them to maintain their false position, and, accept the correct Archimedean analysis as part of official natural philosophy while maintaining Aristotelian prestige in developing experimental physics. On several other matters about that time, to Galileo's amusement, professors of philosophy did discover that Aristotle had not really meant what it had seemed to them he had said. In my final dialogue I have had Simplicio support Galileo's interpretation of Aristotle's meaning, in order to show how easily philosophers could have avoided outright rejection of Galileo's physics at the beginning. With small revisions, Aristotelian natural philosophy might have remained dominant in early modern physics; small not only absolutely, but also relative to revisions that had already been made during the Middle Ages in order to accommodate theological dogmas or permit the use of mathematics in physics. Instead, Aristotelians decided to stand their ground intransigently. Before coming to the particular event that confirmed them in their error, it is instructive to digress to another matter.

When the Copernican issue came to a head four years later, a remarkably parallel situation arose. On that issue the philosophers could not reinterpret or revise Aristotle, who had unequivocally placed the earth motionless at the center of the universe and had based his natural philosophy on that conception. Theologians, on the other hand, faced no equivalent difficulty. The Bible said little about the fixity of the earth, and what it said was inconsistent. What it said about motion of the sun could be construed as metaphorical rather than literal; for example, "the sun goeth forth as a bridegroom from his chamber." Even "the sun also riseth and the sun goeth down," as Galileo pointed out, conformed to the words of Copernicus, when he went right on speaking of sunrise and sunset. But instead of taking the noncommittal course and leaving the question open, the theologians responsible for determining the status of debated doctrines supported Aristotelian philosophy intransigently, just as professors of philosophy had done in 1611, basing their ruling on a declaration that the Copernican propositions were "absurd and foolish in Philosophy" and *therefore* erroneous in the Catholic faith.

I do not know why the responsible theologians backed the philosophers instead of allowing metaphorical interpretation of the Bible. Why philosophers did not reexamine their first hasty interpretation of Aristotle on floating bodies, on the other hand, is both clear and interesting. In the original debate at Salviati's palazzo they had simply overlooked certain cases of floating that appeared to contradict the principle of Archimedes adopted by Galileo. This was called to their attention three days later by Lodovico delle Colombe, who offered to show that bodies denser than water could float in one shape but not in another. Heartened by his experimental evidence, the professors decided to fight Galileo unyieldingly.

It was a poor decision to fight a scientific battle unyieldingly, though that is the proper way to fight philosophical battles. In the latter, Aristotelian professors had had a long history of triumphs, but not by making experimental evidence the test of correctness. Their strength was in logic, and their successes had been gained by substituting logic for experiment, an invincible strategy except in physical science. In science, as Galileo wrote, "a thousand Demostheneses and a thousand Aristotles would be left in the lurch by everyone of average intelligence who happened to hit on the truth by himself" (*Dial.,* p. 54). Whatever shows up in experiments counts, no matter how unyieldingly a logical position may be defended.

Galileo, who more carefully examined the kind of floating introduced by Colombe, found that it accorded with the principle of Archimedes in a very interesting way, and this in turn enabled him to devise further experiments that destroyed conclusions drawn by his adversaries. Meanwhile, made overconfident by Colombe's qualitative and unexamined experiment, they went on ramifying the implications of their position until great sections of Aristotelian natural philosophy were made to depend on a single minor phenomenon of nature. That was their customary practice, and it had been the source of both the strength and the weakness of Aristotelianism. Its strength came from the power of explaining all phenomena in terms of unquestionable metaphysical principles. Its weakness became evident when a series of varied conclusions were separately seen to be incompatible with experience. One was dependence of speeds in fall on weight, shown false by Stevin, if not also by Galileo at Pisa. Another was the celestial location of new stars, established by Tycho Brahe in 1572 and confirmed by Kepler and Galileo in 1604. In 1610 Galileo measured mountains on the moon and found new moving bodies in the heavens; later the same year he observed the phases of Venus. All these things contradicted Aristotelian assumptions. By the time of the 1611 debate, Galileo knew that Aristotelian natural philosophy would collapse if it refused to accept new experiences. He did his best to warn philosophers about making unnecessary commitments they would come to regret, much as he did his best a few years later to warn theologians against gambling the Christian faith on a fixed sun at the center of the universe. But philosophers preferred Aristotle, and so did theologians.

THE FIRST DAY

mathematics the language of nature

Strictly speaking, Galileo regarded mathematics as the language of natural philosophy (physical science) rather than as the language of nature herself, since the "universe open to our gaze" is not literally composed of geometric figures. What he wrote in 1623 was: "Philosophy is written in this grand book, the universe, which stands continually open to our gaze. But the book cannot be understood unless one first learns to comprehend the language and to read the letters in which it is composed. It is written in the language of mathematics, and its characters are triangles, circles, and other geometric figures without which it is impossible to understand a single word of it" (*D&O,* pp. 237–38).

The "book of nature" metaphor was very popular at this time, not only in Italy but also in England, where Shakespeare was writing of books in running brooks and sermons in stones. Confusion of nature with natural philosophy was not anticipated by Galileo when he wrote the above lines; he did not mistake trees for triangles.

"paradoxes" in Galileo's *Discourse*

Letters informed Galileo that readers in various places who at first thought his book paradoxical later came to accept it. The main basis of acceptance was not preference for Galileo's principles but replication of his experiments. The initial reaction of Cremonini was reflected in a letter written in June 1612 to Galileo by his friend Paolo Gualdo at Padua: "Your book on water is in the hands of all these philosophers, but they are silent and do not want to talk about it except to say that one would have to try the things you discuss so subtly, many of which they will not accept without proof. In fact, you have administered to them medicines that give them cramps" (xi, 346). Mathematical demonstrations were not regarded by philosophy professors as proofs in physics; they apparently would accept experiments, but did not deign to try those Galileo described.

Laymen reacted differently. Mark Welser, duumvir of Augsburg, wrote to Paolo Gualdo in July 1612: "I have begun reading it, and so far as I see up to now, it is to me a beautiful, curious, and useful work that will newly swamp philosophers of the ordinary school and will give us things to do and say. But truth will prevail; and for heaven's sake let us not wrong our age by preferring ancient errors to truth newly discovered" (xi, 360). The painter Ludovico Cigoli wrote from Rome to tell Galileo not to worry because his book displeased philosophers: "I think the same happened when Michelangelo began doing architecture outside the rules of others up to his time, whereupon they all put their heads together and said that he had ruined architecture by departing from Vitruvius" (xi, 361). G. B. Agucchi also wrote from Rome to say that the first thing disputed was Galileo's saying that water increases in volume when frozen, the adversaries having appealed to the authority of Hippocrates, whereas Agucchi asked Galileo for clear experimental evidence.

sensate experiences and necessary demonstrations

Soon after Salviati left for Spain in 1613, a professor of philosophy at Pisa opened attacks against Galileo on the ground that motion of the earth contradicted the Bible. Galileo then wrote at length on the place of scripture in purely physical debates, which he said should be settled by sensate experiences and necessary demonstrations that would then reveal the true meaning of scripture, that being written in ordinary language for the use of everyone (*GAW,* pp. 224–29). Galileo would have restricted science to matters capable of being established by experience and proof, leaving all other matters to philosophy, theology, and the humanities. This proposal was greatly expanded in 1615 (*D&O,* pp. 175–216); it was not accepted by theologians, or indeed by most scientists, for some two centuries, and is still contested by philosophers.

mathematicians and philosophers

Sagredo in the flesh was actually ahead of Galileo in adopting the position cited in the first of these notes, above. In August 1612 he wrote to Galileo that he himself spoke of physical matters only "by negation"—that is, by elimination of the untenable, much in the manner Sir Karl Popper advocates today. "In theorizing," Sagredo wrote, "it seems that geometry is incorporated with physics; nevertheless [in optics] I wish to theorize with the intention of assuming physical propositions, or with physics mixed in, which are patent to the senses, and afterward to speculate within sure geometric bounds and experiences, coming thus to knowledge of truth" (xi, 371). Galileo's reply is lost, but his misgivings about Sagredo's procedure are discernible from a later letter of Sagredo in August: "And, although in my letter to you I distinguished philosophers from mathematicians (at which you showed yourself somewhat scandalized), I should like you to know that I use those two terms in accordance with the common interpretation of ordinary people, who call 'philosophers' men who, understanding nothing of physical things (and being even incapable of understanding them) make profession of being nature's own secretaries, and with that reputation pretend to stupefy all the senses of mankind and to deprive men also of the use of reason" (xi, 379).

The position assigned to Sagredo in the dialogue was properly Galileo's at this period. At least one eminent mathematician, Luca Valerio, likewise regarded philosophy as empty verbalism (*GAW,* pp. 190–91).

"The Calculator" was Richard Swineshead (fl. 1350), who was

mentioned in Galileo's earliest writings at Pisa. A pioneer in the mathematics of uniformly accelerated motion, he attempted no actual measurements and was content to solve abstract problems.

origin of the dispute over floating bodies

This is well documented by letters, notes, Galileo's unpublished treatise addresed to Cosimo, and the reply to the *Discourse* by Lodovico delle Colombe (cf. *GS,* chap. 8, and vol. iv of the *Opere*).

prefatory section of the *Discourse*

What Galileo's public expected from him was a book on the system of the world, promised in his *Starry Messenger* of 1610. It was never written, Galileo having been "stayed by a higher hand" in 1616. Translators of his later *Dialogue* mistook that for the book and supplied the title *Systema Cosmicum* in 1635 and *The Systeme of the World* in 1661, though Galileo's own title for it was *Dialogue on the Tides.* This was deleted by the censor at Rome in 1630, leaving only a long subtitle mentioning the two chief world systems.

Galileo's telescope could not resolve the rings of Saturn; that planet appeared to him as having two "ears," whence his term "three-bodied." The data he gave for the periods and hourly motions of Jupiter's satellites were very nearly correct, which made it easy for others to publish complete tables of the motions very similar to those Galileo was already using. For the history of Galileo's determination of the satellite motions, see my article "Galileo and Satellite Prediction," *Journal for the History of Astronomy* 10, 2 (1979): 75–85. The device mentioned by which he measured very small angles telescopically is described in S. Drake and C. T. Kowal, "Galileo's Sighting of Neptune," *Scientific American* 243, no. 6 (December 1980): 74–81.

Sunspots had been shown to others by Galileo at Rome in April 1611, and he conducted careful observations of them while polishing the *Discourse* for publication. A consequence of such observations for motion of the earth was later said by him to have been discussed with Salviati (*Dial.,* p. 345), but other evidence shows that the argument occurred to him much later (*GAW,* pp. 311, 332–35). The main consequence he had in mind in 1612 was that the heavens were far more tenuous than air and were not composed of solid crystal as philosophers supposed: "I think this circumambient substance to be very fluid and yielding—a proposition that appears most novel in the ordinary philosophy" (*D&O,* p. 112). I have accordingly permitted Salviati to hint that Copernicanism was one of Galileo's conclusions in 1613 because his one published unequivocal statement favoring Copernicus appeared at the end of his *Sunspot Letters* (*D&O,* p. 144).

reliability of mathematical rules

In his argument with Cremonini over the location of the new star of 1604, Galileo extended the rules for surveying to astronomical distances by an assumption that Cremonini held to be unwarranted (*GAP,* pp. 11–13). In the particular matter under debate Galileo was justified, but three centuries later it was found that a similar simple extension of ordinary mathematical rules could not be made for very high speeds. Galileo himself realized that the parabolic path of projectiles near the earth's surface could not be extended much beyond that (*GAW,* pp. 378–79; *TNS,* pp. 222–24). I have allowed Simplicio to take credit for Cremonini's cautionary restriction against arbitrary extension of mathematical rules in physics.

Eudoxus as founder of astronomy

Eudoxus created the system of homocentric spheres that Aristotle adopted with some modifications for his cosmology, but that was hardly an "astronomy." Purely mathematical and schematic, it accounted for the observed stoppings and turnings of planets as seen from the earth, but it did not provide means of predicting even eclipses of sun and moon. For those it was necessary to move the earth from the exact center of all celestial orbs, against the philosophers. The compromise adopted about 150 B.C. was stated by Geminus (Thomas Heath, *Aristarchus of Samos* [London, 1913], pp. 275–76). Astronomy was separated from physics and treated as merely technical mathematics serving practical purposes *(techne),* while science *(episteme)* with its causal explanations was reserved to natural philosophers.

moment

This key term in Galileo's physics is widely misunderstood. It originated among mathematicians investigating centers of gravity, as a term denoting in effect the product of a weight and the length of a lever arm. That is still called *moment;* Galileo added to the purely static concept the idea of a very small (or "virtual") movement, foreshadowing the concept of momentum as the product of a mass (for Galileo, a weight) and a velocity (for Galileo, a speed). The medieval concept of *impetus* as an impressed force was abandoned by Galileo in favor of impressed motion, to the speed of which he applied the word *impetus* in his later books without any implication of force. In rejecting force from his physics, Galileo followed Aristotle in preference to his medieval successors, having found ways to measure speeds but not forces. Accordingly, the science of motion that Galileo developed mathematically was what we call *kinematics* (or *kinetics*) and did not include dynamics, the science of forces.

In his notes to the book published by the "Unknown Academician" (Papazzoni) against the *Discourse,* Galileo replied to the professor's objection that the word *moment* was not in the dictionary: "In your language is not used, not just this word *moment,* but not even any of the other commonest words in the whole of mathematics" (iv, 158, n. 16). It was such ignorance that Galileo had in mind when he wrote in 1623 that without the language of mathematics no one could understand anything of the universe that lies open to our eyes. The word *moment* was duly added to the famous Italian dictionary of the Crusca Academy in that same year, when the second edition was published, and Galileo was added to the list of authors used as authorities on the Italian language.

Archimedes' first postulate

The law of the lever, basic to all mechanics, was derived by Archimedes from several postulates, of which the first was: "Equal weights, suspended at equal distances from a center, will remain in equilibrium." Like the physical postulate on which Archimedes founded hydrostatics, this was based not on experiment but on what may be called the negative principle of sufficient reason. No one could deny it because its truth is, so to speak, built into our language itself. Sagredo's discussion of it here expresses my own views, not founded on contemporary documents.

When Galileo derived the law of the lever, he adopted the above postulate alone and proceeded differently from Archimedes (*TNS,* pp. 110–13, adapted from Galileo's unpublished treatise on mechanics as revised about 1601; cf. *MM,* pp. 153–57).

nature must obey our arguments

Galileo had Sagredo say later: "It always seems to me to be extreme rashness on the part of some when they want to make human abilities the measure of what nature can do" (*Dial.,* p. 101), and "To me, great ineptitude exists on the part of those who would have it that God made the universe more in proportion to the small capacity of their reason than to His immense, His infinite power (*Dial.,* p. 370).

moveable

The unusual spelling is adopted for Galileo's noun *mobile* throughout his text, because we have no precise modern equivalent. In the Middle Ages there had been a long and vigorous controversy over the propriety of speaking in natural philosophy (or physics) about body as such in discussing problems of motion. Aquinas took the position, against his teacher Albertus Magnus, that the properties of body belonged not to physics but to metaphysics. Galileo was careful not to say "movable body," as Descartes and all physicists after him began to do. Hence his *mobile* meant "movable entity" of whatever kind, for which we have no precise term and for which I have adopted a legitimate though uncommon spelling.

Aristotle in his *Questions of Mechanics*

This, the oldest extant treatise on mechanics, was probably written by an early follower of Aristotle. The principle to which Galileo alluded was one he used when he added the idea of a small movement to the static concept of *moment* and thus foreshadowed the concept of *momentum.* When revising his earliest discussion of motions along inclined planes (about 1602), Galileo explicitly restricted the assumed movement to the instant of the beginning of motion from rest, as was done later in the "principle of virtual velocities," a very powerful tool of kinematics (*MM,* pp. 63–65, 171–75). An application of this will be seen later in Galileo's addition to the second edition explaining the ability of a small weight of water in a narrow tube to balance the great weight of water in a connected thick cylinder.

Simon Stevin

After the works of Archimedes were printed in the mid-sixteenth century, his admirers frequently objected to the use of any reasoning from motion in matters of equilibrium. The Marquis Guidobaldo del Monte, Galileo patron and the author of the first important text on mechanics, was held back from proposing conservation laws like those later advanced by Galileo on account of his conviction that an unbridgeable gap separates equilibrium from motion (*Mech.,* pp. 300, 308, 316). Stevin offered a syllogistic argument against any attempt to bridge it, amusingly adopting Aristotelian logic to defeat Aristotelian emphasis on motion:

> That which stands still does not describe a circle;
> Two heavy things of equal apparent weight hang still;
> Therefore they do not describe circles.

This was aimed by Stevin against the Aristotelian author of the *Questions of Mechanics,* from whom Galileo nevertheless adopted the principle that enabled him to go beyond Archimedes in both mechanics and hydrostatics. Cf. *The Principal Works of Simon Stevin,* ed. E. J. Dijksterhuis, vol. 1 (Amsterdam, 1955), p. 509. Whether Galileo or his friends knew Stevin's work is debatable. Since it had been published in Latin translation in 1605, there is no reason they should not have read it. On the other hand, technical mathematical books were little in demand at Italian universities then, and after Galileo moved to the Florentine court in 1610 he probably had even less access to such books published abroad than when he taught at Padua.

machine in nature

Normally a weight cannot raise a greater weight without the use of some kind of machine—lever, pulleys, windlass, or the like. It was thus an apparent paradox that a few pounds of water should raise a much heavier beam. This "hydrostatic paradox" had been discussed in other forms by Benedetti and Stevin; Galileo appears to have been the first to analyze this form of it.

a world on paper

Kepler had used the same phrase in a letter to Galileo, contrasting the cosmologies of philosophers with astronomies based on careful observations and measurements. Galileo applied it to distinguish verbal worlds of philosophers from the sensible world that alone was the subject of his dialogues (*Dial.,* p. 113). The distinction Galileo frequently made between science and philosophy has escaped most British scholars and some of their American followers, because "natural philosophy" remained Newton's phrase for physical science and continued in vogue in England until the turn of this century. The word *scienza* (knowledge) had long existed in Italian and was not interchangeable with *filosofia* so far as Galileo was concerned. In English, the word *scientist* did not appear until 1840. Many English writers still call Galileo a "philosopher" simply because he was a physicist, though historians of philosophy ignored Galileo until very recently.

ships in harbors near rivers

In the Fourth Day, Galileo comments on the Aristotelian idea that depth of water affects its buoyancy. From his remarks there about fresh and salt water, I suppose he would have accounted for Aristotle's mistake in the way I have had Salviati say he did.

THE SECOND DAY

esplanade on the Arno

The event was recounted in Galileo's apologetic treatise addressed to the grand duke, from which he then canceled it (iv, 32, n. 2). Don Giovanni de' Medici was not mentioned, but other evidence suggests that he was the offended engineer. Some say he was offended by Galileo earlier, when teaching at Pisa, but Don Giovanni was not in Italy at that period. In 1611 he sided with Galileo's adversaries and Colombe dedicated his attack against the *Discourse* to him (iv, [315]). His antagonism probably originated in 1608, when he and Galileo were both in Florence for Cosimo's wedding.

paradoxes in the *Discourse*

Apart from the genuine hydrostatic paradox, previously discussed, it seemed to contemporary readers that the floating of chips denser than water violated the principle of Archimedes, so that Galileo's reconciliation of the two was paradoxical. Mark Welser wrote to Galileo in October 1612 concerning the *Discourse:* "I have managed to see another copy [of the second edition?], the perusal of which has so won me over (and I do not blush to confess this) that though at the outset your position appeared to me most paradoxical, I now find it to be beyond question. It is so well provided with reasons and experiments that I certainly fail to see how and where your adversaries are going to assail it, though I suppose they cannot be very happy about it" (*D&O,* p. 122).

Galileo replied on 1 December: "Your Excellency remarks that at your first reading of my tract on floating it appeared to you paradoxical, but that in the end the conclusions were seen to be true and clearly demonstrated. You will be pleased to know that the same has happened here to many persons who have the reputation of good judgment and sound reasoning. There remain in opposition to my work some stern defenders of every minute Peripatetic point. So far as I can see, their education consisted in being nourished from infancy on the opinion that philosophizing is and can be nothing but the making of a comprehensive survey of Aristotle's texts, so that from divers passages they may collect and put together any number of solutions to any proposed problem. They wish never to raise their eyes from those pages—as if this great book of the universe had been written to be read by nobody but Aristotle, and his eyes had been destined to see for all posterity" (*D&O,* pp. 126–27).

Francesco Buonamico

Professor of philosophy when Galileo was a student at Pisa, Buonamico published an enormous folio on motion, entirely philosophical and nonmathematical, even ignoring medieval mathematical physics. His philosophical interpretation of Archimedes' *On Bodies in Water,* a book exclusively mathematical and devoid of any metaphysical theory of matter or motion, shows how far physics remained from mathematics in Italian universities of the time.

science and causes

The position here placed in Simplicio's mouth as spokesman for Aristotelianism—that understanding of causes is all that counts in science—was no less that of the staunch *anti*-Aristotelian René Descartes, who condemned Galileo's sciences as built on air because he did not start by revealing the cause of motion (*GAW,* pp. 387–93). It remained the position of philosophers long after scientists had stopped bothering about it, though even scientists were slow to adopt a position as radical as that of Sagredo or Galileo, in which metaphysics and theology were to be left strictly to philosophers and theologians. "Such profound contemplations belong to doctrines much higher than ours, and we must be content to remain the less worthy artificers who discover and extract from quarries that marble in which industrious sculptors later cause marvelous figures to appear that were lying hidden under those rough and formless exteriors" (*TNS,* pp. 182–83).

authority and science

Galileo's implication that Archimedes had no more *authority* in science than Aristotle deserves attention. A few pages later it is said: "the Archimedean doctrine was true because it closely fitted true experiences." Although this was said in a conditional context, Galileo's view that experience was the ultimate authority in science is clear. He did not prefer the teaching of Archimedes because that was rigorously mathematical, or at any rate he did not say so, for Galileo knew that many rigorously mathematical conclusions had no counterparts in the sensible world (*TNS,* pp. 223–24).

treatise on motion composed at Pisa

This treatise, *De motu,* drafted in various versions from 1587 to 1592, survives in manuscript (*Mech.,* pp. 331–77; *MM,* pp. 13–131). Concerning the point made by Salviati, see S. Drake, "The Evolution of Galileo's *De motu,*" *Isis* 66 (1975): 239–50.

cause present or absent with effect

Galileo's definition of cause is found in his notes early in the dispute: *Causa è quella, la qual posta, sèguita l'effetto; e rimossa, si rimuove l'effetto* (iv, 21 [19–20]): "Cause is that which put [placed], the effect follows; and removed, the effect is removed." This definition does not appear in the *Discourse* until well into the Third Day, a page before Galileo's proof that bodies weigh equally when their volumes are inverse to their specific weights. It amounts to substituting a correlation between two sets of appearances for the philosophical concept of cause as an invisible power compelling one appearance to follow another. Galileo's approach in his *De motu,* twenty years earlier, was traditionally causal; in his *Dialogue,* twenty years later, the word *cause* seldom appeared, and in his final *Two New Sciences* Galileo considered the cause of acceleration in fall irrelevant to science and called all philosophical suggestions about it "fantasies" (*TNS,* pp. 188–89).

The move from pursuit of causal inquiries to impatience with them was perhaps the most revolutionary aspect of Galileo's new sciences. It was a gradual move, because it would take a man who was brought up in the old tradition most of a lifetime to see that causes had not been discovered, but were invented arbitrarily by philosophers, and that a great deal more had to be known about phenomena before it could be profitable to consider what might lie behind them. By the time science matured in the nineteenth century, scientists generally became as impatient with metaphysics as Galileo had been.

science will enable men to philosophize better

"Philosophy cannot but benefit from our disputes, for if our conceptions prove true, new achievements [in it] will be made; if false, their rebuttal will further confirm the original doctrines. Save your concern for certain philosophers; come to their aid and defend them [Simplicio]. As to science itself, it can only improve" (*Dial.,* pp. 37–38). When Simplicio went on arguing from Aristotelian principles that change in the heavens was impossible, Salviati replied: "It would be an easier thing to determine whether the earth, a most vast body and very convenient to us because of its proximity, moves so rapidly as to rotate on its axis every twenty-four hours, than it would be to understand and determine whether generation and corruption arise from contraries, or indeed whether generation, corruption, and contraries have any place in nature" (*Dial.,* p. 39). This throws light on Galileo's conception of science as a method capable of human pursuit in contrast with philosophy as a chimerical promise of great things cited in the first note above.

objections by Professor di Grazia

Induction from observations was approved by Aristotle, so Galileo had hope of winning natural philosophers to his side at this stage of his career. Design of experiments, however, had never been discussed; only the deductive logic of causation was outlined by Aristotle in his *Posterior Analytics.* Galileo's reasoning about selection of material for determining the role of shape in floating, sinking, or rising in water is followed later in the *Discourse* by other experimental procedures remarkably similar to ours, rules for which were given by John Stuart Mill and are known as "Mill's canons." Vincenzio di Grazia attacked Galileo's procedures in his book, to which replies were published in 1615 by Benedetto Castelli based on Galileo's notes. (Cf. *GAW,* pp. 220–21; for full texts see iv, 377–440, 696–707).

I have carried out the experiments

Sagredo here speaks for the translator, who has duplicated the experiments described in the *Discourse* and whose results are recounted in the dialogue by Sagredo. Galileo specifically ascribed to him only long and unsuccessful attempts to weight a wax ball with grains of sand so that it would remain motionless in water wherever placed (*TNS,* p. 72). This appears to have been about 1603–4; see note concerning salt water laid beneath fresh water (first note to Third Day).

challenge by Colombe

After the initial discussion between Galileo, Coresio, and di Grazia, the principal focus of attention in the debates of 1611 became the floating of lamina denser than water. Colombe nursed an old grudge against Galileo relating to a different controversy (*GAP,* pp. 133–36). Floating of flat objects had been mentioned by Aristotle, though he did not specifically assign shape as its cause and turned instead to the supposed resistance of water to division. Colombe not only claimed shape as the cause but believed that was Aristotle's opinion; Galileo's arguments against that are found in the Fourth Day. At first he regarded Colombe's intervention as irrelevant to the dispute, which concerned ordinary floating of bodies placed in water with no special precautions, or their rising or sinking. Soon afterward he agreed to debate with Colombe and to reconcile anomalous floating with the principle of Archimedes, but the grand duke advised him instead to write out his arguments and to avoid oral controversy.

In preparation for the debate (which was never held), Galileo wrote various notes that are now bound in scattered form together with surviving parts of the unpublished apologetic essay addressed to Cosimo (iv, 18–29). These notes appear to have begun with a definition of *cause* under which shape as such could not be the cause of anomalous floating. The notes throw light on an intermediate stage of Galileo's reasoning that is of interest with regard to his first approach to the problem and explains his later emphasis on the role of air, on wet and dry surfaces, and on the effort required to separate closely fitting surfaces. Anticipating a challenge to adduce some other cause of anomalous floating when he rejected shape, Galileo reasoned that the action of shape was to alter the area of that part of the floating object that remains above water and in contact with air. Since absence of such contact both implies and is implied by sinking, while its presence implies and is implied by floating, this supplied a cause, under Galileo's definition, that was not shape itself but a consequence of shape and that was the immediate cause of floating which Galileo regarded as demanded by science in preference to mediate or remote causes.

Galileo was led to this approach by observing that, whereas in normal floating a part of the floating object always projects above the water surface, a body denser than water floats entirely below the surface of the surrounding water. This supplied him with an analysis of anomalous floating in which the Archimedean principle not only was vindicated, but offered a means of predicting observable consequences of which some were measurable. In developing these, Galileo had also to explain why water failed to run over the object, for which he appealed to the commonly observed fact that closely fitting surfaces resist immediate separation. This "adhesion of air to dry surfaces" was interpreted by an adversary as the assertion of a "magnetic force of air."

Galileo's exact words

Many notions are still ascribed to Galileo for which direct evidence cannot be found in his precise words but which depend on preconceptions left unexamined by those who assert them. The acuteness of Galileo's refutations of opponents leaves little doubt that he examined his own assertions with equal care and that they conform to some standpoint from which they appeared to him consistent and in agreement with his observations. It has become popular to impute internal contradictions to him and to account for those by supposing him to hold mystical views and an unreasoning zeal for certain ideas, notably that of the Copernican system. His approach to science can be profitably studied in hydrostatics by reason of its relative freedom from mystical elements and from connection with Copernicanism.

mother philosophy as a guide

Replying to Simplicio's fear that men would be lost without Aristotle as their guide in natural philosophy, Salviati said in Galileo's later *Dialogue:* "There is no danger that such a multitude of great, subtle, and wise philosophers will allow themselves to be overcome by one or two [of us] who bluster a bit. Rather, without even turning their pens against those, by means of silence alone, they place them in universal scorn and derision. It is vanity to imagine that one can introduce a new philosophy by refuting this or that author. It is necessary first to teach the reform of the human mind and to render it capable of distinguishing truth from falsehood, which only God can do" (*Dial.,* pp. 56–57). "We need guides in forests and in unknown lands, but on plains and in open places only the blind need guides. It is better for such people to stay at home, but anyone with eyes in his head and his wits about him could serve as a guide for them" (*Dial.,* p. 112). Galileo regarded philosophy as a forest or unknown land and science as a plain or an open place, surrounded by forests of philosophy but accessible to independent exploration.

truth reveals herself tranquilly

"Those who through carelessness are induced to support error will shout loudest and make themselves more heard in public places than those through whom truth speaks, which unmasks itself tranquilly and quietly, though slowly" (iv, 31).

"total sinking" and "partial sinking"

I have placed this terminology in Salviati's mouth to epitomize Galileo's discovery that, although in normal floating there is projection above the general water level, in anomalous floating there is not, so that a kind of continuity exists between the two that is capable of mathematical description and analysis. In contrast, two of Galileo's opponents introduced "total division of water" and "partial division," capable only of arbitrary assertion and ad hoc support (cf. iv, 573–74; also iv, 68, 168–69).

ignorance had been his best teacher

Vincenzio Viviani, Galileo's "last pupil" and first biographer, wrote: "In the matter of the dispute that arose about floating, Galileo used to say that there had been no subtler and more industrious teacher than ignorance, for through that means he had come forth with many ingenious conclusions, and with new and exact experiments to confirm them to the satisfaction of his adversaries' ignorance, to which [experiments] he would not have applied himself to appease his own mind" (xix, 613). This has sometimes been mistranslated and used to support a contention that Galileo denied the need of any experiments whatever, a view hardly compatible with his writings on hydrostatics.

some determinate slowness

It was customary to speak of "degrees of slowness (tardity)" as well as "degrees of speed (velocity)," whereas we now speak only of speeds and regard "zero speed" or rest as the lower limit. In his studies of acceleration from rest, Galileo found it useful to treat rest as "infinite slowness" because of certain perplexities that arise in treating zero and finite magnitudes as mathematically continuous. Galileo's terminology in the present debate showed his opponents that on Aristotle's own principles they could never reach absolute rest by simple increase of slowness, because Aristotle denied any actual infinite in nature. Galileo regarded resistance to division on the part of water as nil, since resistance was to speed of motion alone.

the leaf has penetrated the water

The term *penetrated* may seem to imply "through," but Galileo meant only "into." If we regard the initial water surface as a fixed plane, it is "divided"; if we regard the water surface as deformable from a plane, it is simply depressed and remains undivided.

water should flood over the chip

Aristotelian natural philosophy demanded that ethical considerations must prevail and that science must show not only how things are in nature, but that it is best that they should be so. During the disputes after publication of the *Discourse,* it was demanded that Galileo explain why nature permitted such an injustice as to allow water to be vanquished by air. Since it was a common fact of observation that a drop will stand alone, and need not flatten itself out, Galileo did not trouble himself about the existence of ridges of water no higher than a large drop. For his reply to the ethical question, see *GAW,* pp. 218–19.

inverted tumbler pressed under water

This experiment is well worth the trouble of repeating. The rising of the object is now explained in terms of surface tension, as anomalous floating itself is explained. Galileo explained it by the presence of air as part of the entire floating object. Under his definition of "cause," one explanation is as good as the other, since something had been found that was invariably present with the effect and absent in its absence. It was Galileo's positivistic concept of "cause" that did not satisfy philosophers (and still does not). Even David Hume, whose view of what we learn from observation alone was like Galileo's, regarded the inability of observation to reveal causes as something philosophically significant. Galileo regarded it as implied in the definition of "cause." As he said, "Reason is not needed where our senses reach."

inner essences

"In our speculating we either seek to penetrate the true and internal essences of natural substances, or we content ourselves with knowledge of their properties [*affezione*]. The former I take to be as impossible an enterprise with respect to the closest elemental substances as with more remote celestial things. . . . But if what we wish to fix in our minds is the apprehension of some properties of things, then it seems to me that we need not despair of our ability to acquire this respecting distant bodies just as well as those close at hand, and perhaps in some cases even more precisely in the former than in the latter. . . . These in turn may become the means by which we shall be able to philosophize better about other and more controversial qualities of natural substances" (*D&O,* pp. 123–24). Since Aristotelian natural philosophy concentrated on the discovery of internal essences, Galileo's position was destructive of that entire enterprise. Philosophers ridicule it because Galileo challenged the very purpose of philosophy; many scientists support it for that same reason.

THE THIRD DAY

salt water under fresh

Historically it was Galileo rather than Sagredo who exhibited this, and he did it as a practical joke. Sagredo had spent much time trying to weight a wax ball so that it would stay still in water (*TNS,* p. 72); Galileo showed him a ball motionless in a bowl of water, having laid fresh water over salt water in it. See x, 110; also B. Dibner and S. Drake, *A Letter from Galileo Galilei* (Norwalk, Conn.: Burndy Library, 1967), pp. 52–53.

introduced by Archimedes

The so-called method of exhaustion used by Archimedes to establish a magnitude (or a ratio) by showing it to be neither greater nor less than a certain value appears to have been introduced by Eudoxus, though Salviati would probably have ascribed it to Archimedes. Though the method is cumbersome, it was invaluable to mathematicians before the invention of the infinitesimal calculus. The Archimedean concept of a quantity that becomes and remains less than any previosly assigned value, however small, had been used by Galileo before 1592 to show that a frictionless body resting on a horizontal plane may be set in motion by any force whatever (i, 299–300). This was perhaps the first application to a physical problem of the Archimedean limit concept. For an example of Galileo's use of the method of exhaustion, see *TNS,* pp. 139–41.

an old theory of continued motion

Aristotle paid little attention to forced motions, which by definition were not natural and had no place in physics as the science of nature. Toward the end of his *Physics,* however, Aristotle mentioned disapprovingly an old theory that a projectile is continued in motion by the pressure of air closing in behind it to prevent the existence of a void. The old theory is called *antiperistasis;* Aristotle preferred to consider some of the moving force as having been imparted to the medium carrying the projectile. In the Middle Ages a force, called *impetus,* was supposed to be impressed in the projectile itself. Galileo eliminated the idea of force and asserted that the only thing impressed in the projectile was the motion it had previously shared with the projecting entity. (*Dial.*, pp. 148, 156).

the crowd is divided

Galileo's metaphor is based not on the experience of shoving one's way through a crowd, undergoing physical contact, but on that of effortlessly stepping through a crowd that divides to let one pass. The metaphor is not very good, though in fact we do move less quickly in such circumstances than over an empty field. Galileo took no measurable effort to be required in moving things aside horizontally (disregarding friction, which was negligible in the motion of water). Measurable effort would be felt only in raising things, and that also would be negligible for a solid descending through water except when the solid was quite large in relation to the size of the container of water. Galileo later extended his reasoning to air (*TNS,* pp. 69–74).

in plain Italian

"I wrote it [the *Sunspot Letters*] in the common language because I need to have everyone able to read it, and for that reason I wrote my last little treatise [the *Discourse*] in the same language; and the reason that moves me is my seeing how young men are sent to the university indiscriminately to make themselves into doctors, philosophers, and the like, so that many who apply themselves to those professions are most inept for them, just as others who would be apt are occupied by family cares or in other occupations remote from literature, who then, though as Ruzzante says they have good horse sense, persuade themselves that those miserable pamphlets that contain great new things of logic and philosophy remain way over their heads; and I want them to see that just as Nature has given them, as well as philosophers, eyes to see her works, so she has given them brains capable of perceiving and understanding these" (*GAW,* p. 187).

resistance at the surface of water

Galileo's opponents appear not to have suggested a resistance existing only at the surface of water, nor is there reason to suppose Sagredo would have, but Galileo did touch on the idea in replying to his critics. When Galileo later mentioned the ability of water to sustain itself in large drops, he had Sagredo suppose as an explanation a "tenacity and coherence" between the parts of water; Salviati then argued against any internal cohesion and confessed ignorance of the cause, which he opined might belong to the surrounding air (*TNS,* pp. 73–74). The discussion related to the Aristotelian concept of resistance of water to division; Galileo acknowledged only resistance to *speed* of motion and deduced the existence of a terminal speed for a body falling through any medium, however tenuous (*TNS,* p. 78).

ridges of the greatest height (or depth)

Galileo noted that a maximum height exists beyond which the ridges collapse and water does flow over a chip. In reply to Simplicio's wish to know the height, Salviati said it was about one-eighth of an inch, which is correct for water at room temperature. The figure is found in a note written by Galileo in a book by an opponent: "I will make a platelet of the same length and breadth, and only one-eighth of an inch thick; and though it will divide the water to but a small depth [by the adversaries' rules], it will be impossible for this to remain afloat" (iv, 169).

Galileo's mathematical approach to physics was not algebraic, as is ours, but considered only *ratios* subjected to the Euclidean restriction that both terms must be of the same kind. Some say that Galileo remained ignorant of even fair approximations to what we call physical constants. It is true that those played no part in his physics, which was concerned only with relations. But it cannot be safely inferred that he was unaware of their approximate values, for the statement of which there existed no international units in his day.

new tests reinforce our confidence

Salviati refers to Galileo's concept of "demonstrative advance" in science as applied in argument ex hypothesi favored by him. That type of argument was well known to Simplicio, for Peripatetic logicians had devoted much analysis to it. Strictly speaking, argument by definitions and hypotheses could never establish more than logical consequences; however such consequences were multiplied in number, they left the truth of the hypotheses completely undecided. Simplicio's challenge, based on that fact alone, was equally accepted by Galileo. What was novel in his mathematical physics was the selection of definitions and hypotheses on the basis of actual careful measurements. Aristotle's physics, if put in *ex hypothesi* form, would depend on arbitrary definitions (e.g., "natural" and "forced" motions) and on hypotheses obtained by qualitative induction from observation (e.g., greater speed near end of fall). Medieval physics included some mathematical definitions (e.g., "mediate denomination") and hypotheses (e.g., the "mean speed" postulate) that were not selected on the basis of, or applied to, actual physical measurements. Simplicio knew only those approaches, in which "demonstrative advance" could yield only consequences having no guarantee other than correctness of deduction.

The novel element of measurement somewhat altered the previous situation. Just as the Pythagorean theorem, which can be rigorously deduced from rather simple and generally acceptable axioms of Euclid, appears to us as a remarkable and unexpected relationship when first reached in that way, so did many of Galileo's theorems in physics, reached *ex hypothesi,* appear remarkable in his time. New consequences subject to independent test by actual measurements lent support to the assumed applicability in nature of Galileo's definitions and hypotheses.

This is intended to throw light on the trouble felt by Simplicio, not to be a complete or exact account of Galileo's approach to science. It seemed to Simplicio that Galileo's method of demonstrative advance could be no more than an attempt by science to lift itself by its own bootstraps. Viewed from a strictly logical standpoint, this was true. Salviati and Sagredo, on the other hand, spoke for a large segment of intelligent opinion among educated Italians who were bored with the philosophical world on paper and were interested in any attempt to reach reliable practical information about the sensible world. In astronomy Kepler, working from the measurements made by Tycho Brahe, supplied the needed new planetary hypotheses, while in physics it was Galileo who supplied the necessary new hypotheses for mechanics, hydrostatics, strength of materials, and natural motion. Demonstrative advance in both astronomy and physics led rapidly to confirmation or modification of these in the scientific revolution. Nothing was proved in the sense Simplicio demanded, and perhaps nothing has yet been proved in that sense. The irritation of philosophers is understandable, and it has taken the form of denigration of science starting with that of Galileo. Eternal, incontrovertible truth remains with philosophy, which still sees through the shallow pretences of science just as Simplicio did.

the tiniest possible discrepancy

Galileo had amusingly illustrated this point when his detection of mountains on the moon aroused greater philosophical opposition than any of his other telescopic observations. Heavenly bodies had to be perfectly spherical for Aristotelian metaphysics to survive: "However, these doctors of philosophy never do concede it [the moon] to be less polished than a mirror; they want it to be more so, if that can be imagined, for they deem that only perfect shapes suit perfect bodies. Hence the sphericity of the heavenly globes must be absolute. Otherwise, if they were to concede to me any inequality, even the slightest, I would grasp without scruple for some other, a little greater; for, since perfection consists in indivisibles, a hair spoils it as badly as a mountain" (*Dial.,* p. 80). Galileo, who wrote this, is now said to have believed that all planetary motions are perfectly circular and absolutely uniform, despite appearances to the contrary, by reason of his blind devotion to the authority of Plato.

tiny corrections will be discovered

"There is not a single effect in Nature, not even the least that exists, such that the most ingenious theorists can ever arrive at a complete understanding of it. This vain presumption of understanding everything can have no other basis than never understanding anything" (*Dial.,* p. 101).

"Dealing with science as a method of demonstration and reasoning capable of human pursuit, I hold that the more this partakes of perfection, the smaller the number of propositions it will promise to teach, and fewer yet will it conclusively prove. Consequently the more perfect it is, the less attractive it will be, and the fewer its followers. On the other hand magnificent [book] titles and many grandiose promises attract the natural curiosity of men and hold them forever involved in fallacies and chimeras" (*D&O,* pp. 239–40).

The mechanical philosophy, introduced by René Descartes in a book titled *Principles of Philosophy,* promised complete solution of every problem by means available to anyone within his own head. Cartesian philosophy superseded Galileo's science two years after his death.

the same cone, base down, will drop to the bottom

The effect described by Galileo is easily tested with chocolate chips of the kind used in cookies. They can be floated only point down, in the position best adapted to divide the resistance of water assumed by Galileo's opponents.

see how far our conclusions depart from exact observation

Sagredo's statement may seem anachronistic, since it is through the theory of observational error and the practice of refined means of measurement that we now would support it. Nevertheless, Galileo and his friends already saw science as a never-ending and self-correcting process. "No firm science can be given of such events as heaviness, speed, and shape, which are variable in infinitely many ways. Hence to deal with such matters scientifically, it is necessary to abstract from them; we must find and demonstrate conclusions abstracted from the [material] impediments, in order to make use of them in practice under those limitations that experience will teach us" (*TNS,* p. 225). The practice of precise measurement was advanced by Galileo in both physics and astronomy beyond anything previously dreamed of, even though his achievements now seem very imprecise to us. (See, for example, my articles in *Scientific American* for June 1975 and December 1980.) The theory of observational error was adumbrated in Galileo's analysis of minimum corrections to the reports of various observers of the new star of 1572 (*Dial.*, pp. 281–307).

empty inverted glass

It is my own conjecture about the origin of Galileo's interesting identification of buoyancy of air in water and weight of water in air that is placed in Sagredo's mouth. I think it probable that when Galileo presented the two efforts as (1) pressing a glass in its normal position down into water, and (2) lifting water alone by raising it in an inverted glass, he did so because he was aware of the slight compression of air in an inverted tumbler pushed into water. As Sagredo reported, that is quite small at little depth, but it increases rapidly. In the "magnetic virtue" experiment the object to be lifted lies on the bottom of the container, so that the tumbler is pushed well below the surface. Galileo often disregarded small discrepancies, especially in books written for the public to show them that in the limited domain of direct experience they could see and think as well as philosophers could.

elements of entire new sciences

Galileo's principal lasting contribution to physics was his analysis of falling bodies and projectiles, which he prefaced by saying: "what is more worthwhile, there will be opened a gateway and a road to a large and excellent science of which these labors of ours shall be the elements, [a science] into which minds more piercing than mine shall penetrate to recesses still deeper (*TNS*, p. 190).

THE FOURTH DAY

he implied much more

This is the present assumption of historians and philosophers of science who reason *ex hypothesi* to a concealed metaphysics of Galileo that enabled him to make one or two useful contributions while implying also grotesque errors in physics and astronomy. Since Galileo forbore to offer any metaphysics as a guide to others, he is vulnerable to having one invented for him to suit his critics. None of the conjectural inventions appeal to me, since all neglect

Galileo's express denial that any natural phenomenon will ever be understood by any theorist. However, it is probably true that "every man possesses some metaphysical system, which he has imbibed he knows not how, and credits he knows not why. Its incomprehensibility renders him sensitive to its preservation. It is an unfortunate child whose very idiocy endears it to his feelings" (A. B. Johnson, *The Philosophy of Human Knowledge; or, A Treatise on Language* [New York, 1828], p. 11).

thin plates of iron . . . stop upon water

Aristotle raised the question precisely because shape could *not* be the cause, though that was the reverse of the position taken by the Peripatetics of Galileo's time. See *De caelo,* last chapter, and Galileo's later analysis.

swim in the air by their smallness

The falsity of this is evident from the dust deposited in an attic that remains closed for a very long time.

purpose of imagined experiences

The analysis is mine, not one set forth in the seventeenth century. Experiments "in imagination" constituted a valuable medieval contribution to scientific thought, as careful measurement in physics constituted a valuable Galilean contribution.

Democritus philosophized correctly

Galileo's comment is limited in scope to the particular question in hand. He was certainly not an atomist in the ancient philosophical tradition, a position especially repugnant to Aristotle. See *GAW,* pp. 362–64, and, for an excellent discussion of Galileo's viewpoint, H. E. Le Grand, "Galileo's Matter Theory," in *New Perspectives on Galileo,* ed., R. E. Butts and J. C. Pitt (Dordrecht and Boston, 1978), pp. 197–208.

no other cause

Galileo considered his three kinds of floating to have been reduced to a single cause, the lesser specific weight of the floating object in comparison with water. For the anomalous case, the true floating object was distinguished from the apparent object by inclusion of a layer of air between that and the surface of the surrounding water. Adherence of air to dry surfaces, though also discussed, was not regarded by Galileo as a *cause* of floating; like the "affinity" between closely fitting surfaces and the "magnetic virtue" of air, it was presented as an observed phenomenon of nature.

band of philosophers at Florence

This group was called the "pigeon league" by Galileo and his friends, as a play on the name of Colombe (dove), who appears to have been its instigator and leader. Meeting at the home of the archbishop of Florence on one occasion, they considered inducing a priest to denounce Galileo from the pulpit (xi, 241–42). The archbishop appears to have rebuked the member who suggested this. Three years later a young Dominican firebrand did denounce the Galileists (and all mathematicians) from the pulpit of Santa Maria Novella. Previously, a Platonist professor at Pisa, in Galileo's absence, had declared to his employers that his belief in motion of the earth contradicted the Bible (*D&O*, p. 151). Probably the Catholic church would not have taken that matter up officially without prodding from professors of philosophy, and when a ruling was adopted it was prefaced by the statement that the Copernican motions were "foolish and absurd in Philosophy."

publish his rejoinders

Galileo's replies to philosophers who had attacked his *Discourse* in print, edited and supplemented by Benedetto Castelli, were held up for a year after they were first given to a printer. Pressure against their publication was certainly not political, since Galileo was much favored by the grand duke, and it was not theological; hydrostatics had nothing to do with the Bible. Presumably it was brought by the philosophers leagued against Galileo, or by the University of Pisa on their behalf as professors whose refutation might tend to discredit that institution.

Index

Causa e quella in quale

et rimossa si rimuove

di Piombo va al fondo;

domando

non si può dire che sia la

dola sott'acqua non si

va al fondo; ma ne è l

va al fondo. dico dunq

no la causa del non des

rimossa ne seguita la

senza mutar la forma